FUZZY SETS IN ENGINEERING DESIGN AND CONFIGURATION

INTERNATIONAL SERIES IN INTELLIGENT TECHNOLOGIES

Prof. Dr. Dr. h.c. Hans-Jürgen Zimmermann, Editor
European Laboratory for Intelligent
Techniques Engineering
Aachen, Germany

Other books in the series:

Applied Research in Fuzzy Technology
by Anca L. Ralescu

Analysis and Evaluation of Fuzzy Systems
by Akira Ishikawa and Terry L. Wilson

Fuzzy Logic and Intelligent Systems
edited by Hua Li and Madan Gupta

Fuzzy Set Theory and Advanced Mathematical Applications
edited by Da Ruan

Fuzzy Databases: Principles and Applications
by Frederick E. Petry with Patrick Bose

Distributed Fuzzy Control of Multivariable Systems
by Alexander Gegov

Fuzzy Modelling: Paradigms and Practices
by Witold Pedrycz

Fuzzy Logic Foundations and Industrial Applications
by Da Ruan

FUZZY SETS IN ENGINEERING DESIGN AND CONFIGURATION

EDITED BY

Prof. Dr. Hans-Jürgen Sebastian
Rheinisch-Westfälische Technische Hochschule Aachen
Institut für Wirtschaftswissenschaften
Operations Research

Templergraben 64
52062 Aachen, Germany

Prof. Erik K. Antonsson, Ph.D., P.E.
Engineering Design Research Laboratory
Division of Engineering and Applied Science
California Institute of Technology

1200 East California Boulevard
Pasadena, CA 91125, U.S.A.

KLUWER ACADEMIC PUBLISHERS
Boston/London/Dordrecht

Distributors for North America:
Kluwer Academic Publishers
101 Philip Drive
Assinippi Park
Norwell, Massachusetts 02061 USA

Distributors for all other countries:
Kluwer Academic Publishers Group
Distribution Centre
Post Office Box 322
3300 AH Dordrecht, THE NETHERLANDS

Library of Congress Cataloging-in-Publication Data

A C.I.P. Catalogue record for this book is available
from the Library of Congress.

ISBN-13: 978-1-4612-8636-3 e-ISBN-13: 978-1-4613-1459-2
DOI: 10.1007/978-1-4613-1459-2

Printed on acid-free paper.

CONTENTS

LIST OF FIGURES ix

LIST OF TABLES xiii

CONTRIBUTORS xv

PREFACE xvii

ACKNOWLEDGMENTS xix

1 MODELING IMPRECISION IN ENGINEERING DESIGN
E. K. Antonsson, K. N. Otto and K. L. Wood 1
1 Introduction 2
2 Engineering Design with Imprecision 8
3 Combining Imprecision in Design 10
4 Uncontrollable Variables in Design (Noise) 19
5 Hybrid Uncertainty 26
6 A Brake Design Application 29
7 Conclusion 47

2 MULTIPLE OBJECTIVE DESIGN OPTIMIZATION
A. K. Dhingra 53
1 Introduction 53
2 Problem Statement 53
3 Crisp Approaches for MultiObjective Optimization 55

4 Fuzzy Approaches for MultiObjective Optimization 58
5 Cooperative Fuzzy Games 65
6 Numerical Examples 70
7 Summary 83

3 INTELLIGENT SYSTEMS FOR CONFIGURATION PROBLEMS
H.-J. Sebastian 89
1 Intelligent Decision Support Systems for Design and Configuration Tasks 89
2 Modelling Design and Configuration Problems Including Fuzziness 95
3 Description of Applications 142
4 Introduction of KONWERK - a Modular Design and Configuration Tool-Kit 164
5 Solving Applications Using the Integrated Approach of Knowledge-Based Techniques with Fuzzy Logic and Fuzzy MCDM 181
6 Summary and Conclusions 226

4 MANAGEMENT OF UNCERTAIN INFORMATION IN SIMULTANEOUS ENGINEERING
W.Eversheim, H.-J. Zimmermann A. Roggatz and T. Derichs 233
1 Introduction 234
2 Aspects of Modeling Uncertain Information 235
3 Model for a Description of Uncertainties on a Contents Level 236
4 Methodology for a Context Dependent Description of Uncertainties 238
5 Information Model to Describe Uncertainties within the Context 243
6 Early Pass on of Information by Structuring Information 244
7 Application of the Methodology 245
8 Conclusion 249

5 APPLICATION OF THE FUZZY AHP METHOD FOR ASSESSING ALTERNATIVE PRODUCTION CYCLES

M. Weck, F. Klocke, H. Schell and E. Rüenauver 251

1 Introduction 252

2 Selecting which Assessment Method to Apply 253

3 Methods of Weighting and Assessing Stages of Production 256

4 Assessing Production Cycles 259

5 Example of an Assessment Process 260

6 A METHOD FOR PERSONNEL SELECTION IN CONCURRENT ENGINEERING USING FUZZY SETS

J. Stahl 265

1 Introduction 266

2 Personnel Selection in Concurrent Engineering 266

3 Integrated Personnel Planning in CE 267

4 Hierarchical Structure of KSAOs 268

5 Nominal Criteria 273

6 Metric Criteria 274

7 Determination of Overall Suitability 274

8 Conclusion 275

INDEX 277

LIST OF FIGURES

Chapter 1

1	Flat and Hemispherical Head Air Tank Designs.	4
2	Preferences: Length l, Radius r, Metal volume m, Capacity v.	7
3	Preferences: Outer Radius R_0, Outer Length L_0.	8
4	Preference for l and r induced onto the PVS (m, v).	10
5	Preferences on the DVS: l, r, m, v.	11
6	Preferences on the DVS: R_0, L_0.	12
7	Flat and Hemi Head Tank Trade-Off Strategy Results.	17
8	Design Variable Probability Distributions.	22
9	Applied Pressure Possibility.	24
10	Preferences for the Flat and Hemi Head Tanks.	28
11	Drum Brake System.	30
12	Disk Brake System.	31
13	Drum Brake: Induced Preferences.	37
14	Preference: Torque Performance Specification, $\tilde{\mathcal{T}}^r$.	38
15	Disk Brake: Induced Preferences (Force and Torque).	40
16	Disk Brake: Induced Preferences (Temperature).	42

Chapter 2

1	Path generating planar four-bar mechanism	72
2	Driving torque as a function of input link orientation	74
3	An ACOSS-FOUR flexible space structure	79
4	LOS error versus time for minimum weight design	82
5	LOS error versus time for minimum control effort design	83
6	LOS error for case (a) of fuzzy game theory formulation	84
7	LOS error for case (d) of fuzzy game theory formulation	84

Chapter 3

1	Areas of Application of IDSS	90
2	Relevant Methods for IDSS	91
3	Simple illustrative example of a configuration problem	92
4	An Architecture of an IDSS	94
5	Conceptual Hierarchy - Car	97
6	Conceptual Hierarchy - a general view	98
7	Conceptual Hierarchy - an example	99
8	Conceptual Hierarchies (Hierarchies of Objects described by Attributes)	101
9	Classes - Subclasses	102
10	Example - FRAME TUBE	103
11	Specialization of a Class	104
12	The Couch Example	104
13	Fuzzy Sets for specific linguistic terms	108
14	Fuzzy Set "Fast-Car"	110
15	Example for an Aggregation	112
16	Algorithm to specify a Fuzzy Multi-Criteria Design Model	115
17	optimization-directed perspective	117
18	Subtree VecOp	120
19	Subtree: Optimization in the case of Fuzziness, multi-criteria, several realization of SPEC	121
20	Two Substitute Criteria in the MADM Case	122
21	Modifications of DECON-1	126
22	Basic Algorithm of Sequential Configuration	132
23	A particular Branch and Bound Strategy	139
24	Cabin Layout that was configured with XKL	142
25	Restrictions for cross aisles between type-A doors with two approach paths	143
26	Customer's Requirement for a certain Seat Pitch	144
27	A selection of existing space launch systems	145
28	Specialization Hierarchy of Space Launch Systems	146
29	Part-of Hierarchy of Instance Ariane V	146
30	Part-of Hierarchy for STS-Objects	148
31	Example - Logistic Layout	150
32	Transportation of letters - an example	151

33 Graphical User Interface 152
34 Subgraph to represent a PC 154
35 Europe 158
36 The Conceptual Hierarchy of the RAINS-model (incomplete because of the complexity) 159
37 Representation of objects by frames 160
38 Representation of hierarchical ordered objects 160
39 Representation of Concepts by Slot-Filler notation to represent the attributes 161
39 Concept of EUROPE as an instantiation of Overall-Concept 161
39 Example of a specialization chain 162
40 Requirement Modelling and Basic Optimization Module of KONWERK 169
41 Problems and Solvers 171
42 Linguistic variable "Style" (of object car) 174
42 Linguistic variable *size-engine* (of object car) 175
43 Chen and Huang's scale to represent imprecise linguistic terms 180
44 Fuzzy Set "Comfortable-Distance" 181
44 Fuzzy Set "Large-Seat-Number" related to a business class 182
44 Fuzzy Set "Fast Delivery Time" 182
45 Illustration of the Best-First-Search - Part of the BaB Algorithm 188
46 An algorithm to estimate the start mass 192
47 Part of the goal hierarchy and alternatives (STS's) 195
48 Linguistic Variable "Noise Intensity" 197
49 Specialization Hierarchy of Logisit Objects 198
50 Simplified Compositional Hierarchy 199
51 Membership function of a fuzzy set "preferred-service-time" (mailbox) 201
52 Cost-function $f_k(x^k)$ for a source-region k 211
53 Membership function for "Acceptable-Grid-Deposition" 212
54 Membership function for "Acceptable-Cost" 213
55 Fuzzy Sets to model Emission-level-of-Source-region-k 214
56 An example of the linguistic variable "Deposition in Grid j" 214
57 Membership function to model an objective as fuzzy constraint 216
58 Linguistic variable "Cost-of-Reductin-Emission" 216

59 Architecture of the "Overall Modelling Approach" 219
60 A special requirement model 222
61 Fuzzy constraint "Desired Cost Level" 224
62 Aggregation tree 225

Chapter 4

1 Aspects for Modeling Uncertain Information 235
2 Presentation of uncertainties on a contents level described in EXPRESS-G 238
3 Determining the information-maturity 239
4 EXPRESS-G Model for structuring information using an object-oriented schema 245
5 First step for the use of information-maturity in Simultaneous Engineering 246
6 Second and third step for the use of information-maturity in Simultaneous Engineering 247

Chapter 5

1 Target in assessing alternative production cycles 252
2 Structuring the decision-making problem and deriving a suitable method of assessment 254
3 Extract of the hierarchy in a decision-making problem 255
4 Determining Kv, the number relating to a specific process 258
5 Approach for applying the Buckley fuzzy AHP method 260
6 Cycles for manufacturing a gear shaft 261
7 Result of the assessment of the alternatives 263

Chapter 6

1 Concept for the Integrated Personnel Planning in CE 268
2 Hierarchical Structure of KSAOs 269
3 Example for Membership Functions of Ordinal Criteria 271
4 Derivation of the Membership Functions for the Linguistic Description 272

LIST OF TABLES

Chapter 1

1	Overall Preference Resolution Axioms.	12
2	*Min* Preference Resolution Axioms.	13
3	*Product* Preference Resolution Axioms.	15
4	Uncertainty Integral Restrictions.	20
5	Brake Example: Imprecise Design Variable Data.	35
6	γ-*level* Measure Results for Drum Brake.	44
7	γ-*level* Measure Results for Disk Brake.	45
8	Disk Brake: γ-*level* Measure Results for T_d.	45

Chapter 2

1	Objective function values for single objective optimizations	73
2	Results obtained with six objectives considered simultaneously	74
3	Comparison of single vs multiple objective optimization results	75
4	Objective function values for single objective optimizations	81
5	Design variables for single objective optimizations	81
6	Objective function values using cooperative fuzzy games	81
7	Design variables using cooperative fuzzy games	82

Chapter 3

1	Survey of the basic and extension modules	166
2	A small car database	175
3	Membership values	176
4	A small example database	177
5	Requirements for a Cabin Layout	186
6	Maximal resolution of Graphic Card	204

7	Number of Colours	204
8	Working Memory	204
9	Bus System	205
10	Monitor-Diagonal	205
11	Vertical Frequency	206
12	Maximal resolution	206
13	Composed Attributes	209

Chapter 4

1	Linguistic terms for a description of the changing frequency	241

CONTRIBUTORS

Professor Erik K. Antonsson, Ph.D., P.E.
Engineering Design Research Laboratory
Division of Engineering and Applied Science
California Institute of Technology
Pasadena, CA 91125, U.S.A.

Dipl.-Ing. Dipl.-Wirt. Ing. Thomas Derichs
Institute for Operations Research
RWTH Aachen, University of Technology
52056 Aachen, Germany

Professor Dr.-Ing. Dr. h. c. Walter Eversheim
Laboratory for Machine Tools and Production Engineering
RWTH Aachen, University of Technology
52062 Aachen, Germany

Professor Anoop K. Dhingra, Ph.D.
College of Engineering and Applied Science
University of Wisconsin, Milwaukee
Milwaukee, WI 53201, U.S.A.

Professor Dr.-Ing. Fritz Klocke
Laboratory for Machine Tools and Production Engineering
RWTH Aachen, University of Technology
52056 Aachen, Germany

Professor Kevin N. Otto, Ph.D.
Department of Mechanical Engineering
Massachusetts Institute of Technology
Cambridge, MA 02139, U.S.A.

Dipl.-Ing. Dipl.-Wirt. Ing. Axel Roggatz
Laboratory for Machine Tools and Production Engineering
RWTH Aachen, University of Technology
52056 Aachen, Germany

Dipl.-Ing. Elmar Rüenauver
Laboratory for Machine Tools and Production Engineering
RWTH Aachen, University of Technology
52056 Aachen, Germany

Dipl.-Ing. Hanno Schell
Laboratory for Machine Tools and Production Engineering
RWTH Aachen, University of Technology
52056 Aachen, Germany

Professor Dr. Hans-Jürgen Sebastian
Institute for Operations Research
RWTH Aachen, University of Technology
52056 Aachen, Germany

Dipl.-Ing. Jürgen Stahl
Institute of Industrial Engineering and Ergonomics
RWTH Aachen, University of Technology
52056 Aachen, Germany

Professor Dr.-Ing. Dr.-Ing. E. h. Manfred Weck
Laboratory for Machine Tools and Production Engineering
RWTH Aachen, University of Technology
52056 Aachen, Germany

Professor Kristin L. Wood, Ph.D.
Department of Mechanical Engineering
The University of Texas
Austin, TX 78712, U.S.A.

Professor Dr. Dr. h. c. Hans-Jürgen Zimmermann
Institute for Operations Research
RWTH Aachen, University of Technology
52056 Aachen, Germany

PREFACE

As understanding of the engineering design and configuration processes grows, the recognition that these processes intrinsically involve imprecise information is also growing. This book collects some of the most recent work in the area of representation and manipulation of imprecise information during the synthesis of new designs and selection of configurations. These authors all utilize the mathematics of fuzzy sets to represent information that has not-yet been reduced to precise descriptions, and in most cases also use the mathematics of probability to represent more traditional stochastic uncertainties such as uncontrolled manufacturing variations, etc. These advances form the nucleus of new formal methods to solve design, configuration, and concurrent engineering problems.

Hans-Jürgen Sebastian
Aachen, Germany

Erik K. Antonsson
Pasadena, California

ACKNOWLEDGMENTS

We wish to thank H.-J. Zimmermann for inviting us to write this book. We are also grateful to him for many discussions about this new field Fuzzy Engineering Design which have been very stimulating.

We wish to thank our collaborators in particular: B. Funke, M. Thärigen, K. Müller, S. Järvinen, T. Goudarzi-Pour, and T. Kriese in Aachen who worked in the PROKON project and who elaborated some of the results presented in the book.

We also wish to thank Michael J. Scott for providing invaluable editorial assistance.

Finally, the book would not have been possible without the many contributions and suggestions of Alex Greene of Kluwer Academic Publishers.

1

MODELING IMPRECISION IN ENGINEERING DESIGN

Erik K. Antonsson, Ph.D., P.E.
Kevin N. Otto, Ph.D.*
and Kristin L. Wood, Ph.D.**

Engineering Design Research Laboratory
Division of Engineering and Applied Science
California Institute of Technology
Pasadena, CA 91125, U.S.A.

* *Department of Mechanical Engineering*
Massachusetts Institute of Technology
Cambridge, MA 02139, U.S.A.

** *Department of Mechanical Engineering*
The University of Texas
Austin, TX 78712, U.S.A.

ABSTRACT

The background, terminology, and theory of a technique (called the *Method of Imprecision* or M_oI) to formally represent and manipulate imprecise information in engineering design calculations is reviewed here [3, 29, 30, 33, 36, 37, 38, 39, 43, 56, 57, 58, 59, 60]. The method utilizes fuzzy sets to capture designers' and customers' preferences. Formal methods for including noise, trade-off strategies and design iteration are included.

Two illustrative examples (one of an air tank design, the other the design of a braking system for a passenger vehicle) are presented. The examples emphasize a comparison of different solution alternatives during preliminary engineering design, involving realistic design complexities and decision criteria. Sample numerical and graphical results are given for the applications, including the imprecise input design variables and output performance variables of the design examples. Normalized measures of the importance of, and coupling between, design variables are also presented.

1 INTRODUCTION

Imprecision and vagueness are intrinsic aspects of engineering design. If (at the start of a design process) a proposed solution were neither imprecise nor vague, its description would be precise and it would therefore be a completed design. While (stochastic) uncertainty typically remains in a completed design description (e.g., dimensional tolerances), the nominal desired dimensions are precise. Engineering design is essentially the process of reducing the ***imprecision*** in the description of solution concepts.

The need for a methodology to represent and manipulate imprecision is greatest in the early, preliminary phases of engineering design, where the designer is most unsure of the final dimensions and shape, materials and properties, and performance of the completed design. The "back of the envelope" sketch commonly used as a starting point for solving design problems, will not include precise dimensions or other attributes, but is almost universally described imprecisely (e.g., "This diameter will be ***about*** 2 cm."). Additionally, the most important decisions, those with the greatest effect on overall cost, are made in these early stages [25, 49, 53, 55]. Other references suggest that approximately 70% of the cost of a product is determined during the preliminary design [7, 24, 52].

> "If a major project is truly innovative, you cannot possibly know its exact cost and its exact schedule at the beginning. And if in fact you do know the exact cost and the exact schedule, chances are that the technology is obsolete." [21][1]

Thus, the information available to an engineering designer in the early stages is intrinsically vague, and obtaining precise information upon which to base decisions is usually impossible. Yet the decisions that are made on the basis of this imprecise information are the most critical.

A technique (called the *Method of Imprecision* or M_oI) has been developed by the authors to represent and manipulate this intrinsic imprecision by incorporating designers' and customers' preferences as fuzzy membership functions [3, 29, 30, 33, 36, 37, 38, 39, 43, 56, 57, 58, 59, 60]. As described elsewhere [3, 29, 54] set-based approaches to design can facilitate concurrent engineering. The application of this method will be illustrated by way of two

[1] Joseph G. Gavin, Jr., discussing the design of the lunar module that landed NASA astronauts on the moon on July 20, 1969.

examples. The first is the design of an air tank (to store or transport pressurized air). The second is the design of a braking system for a passenger vehicle.

1.1 Example: Design of Air Tanks

The example presented below considers a pressurized air tank design, and is the same problem as presented in Papalambros and Wilde [40], page 217. Reference [40] describes the restrictions applied to the problem to permit it to be solved using crisp constraints and various optimization techniques (monotonicity analysis, non-linear programming). The example was chosen for the presentation here for its simplicity, and also because its (fuzzy) preferences can be represented on a plane for plotting and visual interpretation.

The design problem is to determine length and radius values for an air tank with two different choices of head configuration: flat or hemispherical, as shown in Figure 1.

1.2 Forming Imprecision

The approach used in the M$_o$I is to elicit preferences for various aspects of the design from the designers and customers. They indicate their preference as membership values (on a scale from 0 to 1) for each value of each variable describing the design.

The customers' needs are described using Performance Variables, which define the desired performance of the design. Each performance variable is denoted p_j, and the associated membership function is denoted $\mu(p_j)$. The Performance Variables span the Performance Variable Space (PVS).

The designers' preferences are described as Design Variables, which include the designers' experience and expertise, as well as availability, manufacturability, and other attributes of the design. Similarly each design variable is denoted d_i, and the associated preference function is denoted $\mu(d_i)$. The Design Variables span the Design Variable Space (DVS).

There are four performance variables in the tank design. The first is the metal volume m:

$$m = 2\pi K_s r^2 l + 2\pi C_h K_h r^3 + \pi K_s^2 r^2 l. \tag{1.1}$$

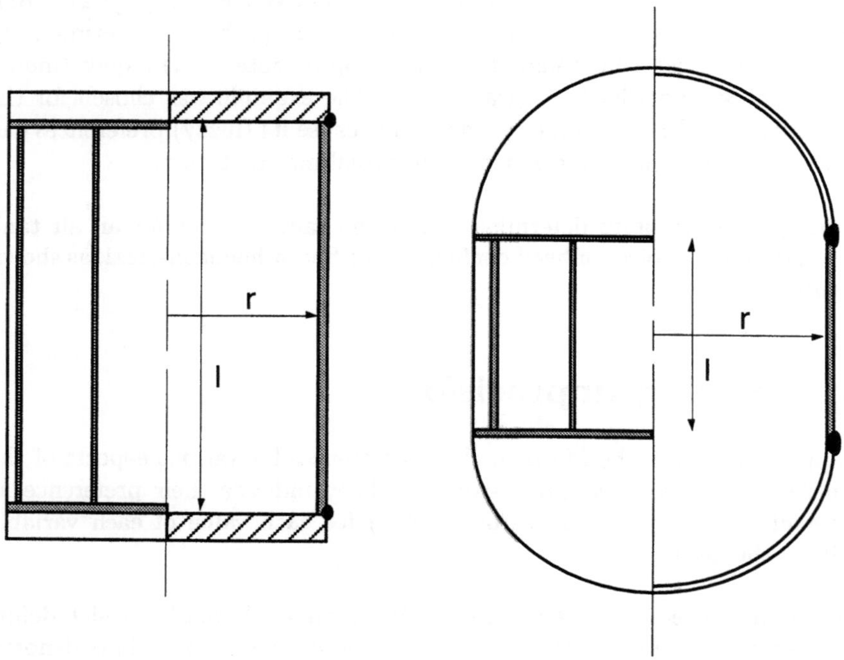

Figure 1 Flat and Hemispherical Head Air Tank Designs.

This variable is proportional to the cost, and the preference ranks are set by the designer based on this concern. Another performance variable is the tank capacity v:

$$v = \pi r^2 l + \pi K_v r^3. \tag{1.2}$$

This variable is an indicator of the design's principle objective, to hold air. Another variable is an overall height restriction L_0, which is imprecise:

$$l + 2(K_l + K_h)r \leq L_0. \tag{1.3}$$

Finally, there is an overall radius restriction R_0, which is also imprecise:

$$(K_s + 1)r \leq R_0. \tag{1.4}$$

The last two performance variables have their preference ranks set by spatial constraints.

The coefficients K are from the ASME code for unfired pressure vessels. S is the maximal allowed stress, P is the atmospheric pressure, E is the joint efficiency, and C_h is the head volume coefficient.

$$K_h = \begin{cases} 2\sqrt{CP/S} & \text{flat} \\ \frac{P}{2S-.2P} & \text{hemi} \end{cases} \tag{1.5}$$

$$K_l = \begin{cases} 0 & \text{flat} \\ 4/3 & \text{hemi} \end{cases} \tag{1.6}$$

$$K_s = \frac{P}{2SE - .6P} \tag{1.7}$$

$$K_v = \begin{cases} 0 & \text{flat} \\ 1 & \text{hemi.} \end{cases} \tag{1.8}$$

This example's design space is formed by the two variables the designer is free to choose, l and r, thus the DVS $\simeq \mathbb{R}^2$. The performance variable space is formed by the four performance variables m, v, L_0, and R_0, thus the PVS $\simeq \mathbb{R}^4$.

The preference information included for each design variable should exclude any information that is modeled by a different variable. For example, information on the cost of the design (and preferences for cost) are included in the preference function for the design variable m above. If a separate performance measure explicitly included cost, the designer should not include considerations related to cost in m's preference function. This may mean that the designer's preference for m will be high (near 1.0) over a large range. The design will be limited by some other aspect that will be included in later design trade-offs.

The ability to include design preferences that are not explicitly modeled is a significant contribution of the MoI over other decision-making methods (e.g., Utility Theory, etc.). Using the MoI, the designer's subjective experience and expertise can be incorporated into design decision-making, including elements of engineering style, component availability, reliability, manufacturability, etc.

A formal method to elicit designer and customer preferences can be implemented by using Utility Theory's "Lottery Question" [9, 17, 18, 19, 26, 27] with the substitution of the word "preference" for "probability".

The preferences for values of the variables in the tank example are shown in Figures 2 and 3 for the hemispherical design; the flat head design space is similar.

The preference information is usually (but not always) a convex fuzzy number. Thus, the MoI uses the mathematics of fuzzy calculus to operate on these imprecise descriptions of the design. However this information is not treated as fuzzy membership, nor are fuzzifying or de-fuzzifying operations performed, nor are logical operators on fuzzy sets performed. Rather, these imprecise preference data are used to perform the usual engineering computations encountered in design to map the preferences of the designers onto the Performance Variable Space, and also to map the preferences of the customers onto the Design Variable Space. Thus *preference* is a map μ_k from a space X_k to $[0,1] \subset \mathbb{R}$,

$$\mu_k : X_k \to [0,1] \tag{1.9}$$

that preserves a designer's preferential order over X_k. This provides the designer with the ability to trade-off the many (and typically incommensurate) aspects of a design in an understandable way.

Therefore, the designer has developed a formal design variable space, DVS, consisting of the alternatives to choose among. The DVS will be characterized by ***design variables***, $d_1, \ldots, d_n$. For a design process of selection among alternatives, each d_i represents an alternative, and so the DVS $\simeq \mathbb{Z}_n$ (the finite set of integers up to n). For determination of values in a design model, there are usually multiple variables each of which could be thought of as a continuum, and so each d_i might be thought of as a vector ($\vec{d_i}$) within a DVS $\simeq \mathbb{R}^n$, where the DVS is represented in some basis with coordinates d^i, for example.

There is also a dependent performance variable space, PVS, consisting of the set of evaluated performances of points in the DVS. The PVS will be characterized by ***performance variables***, $p_1, \ldots, p_m$. Again, each p_j could be the result

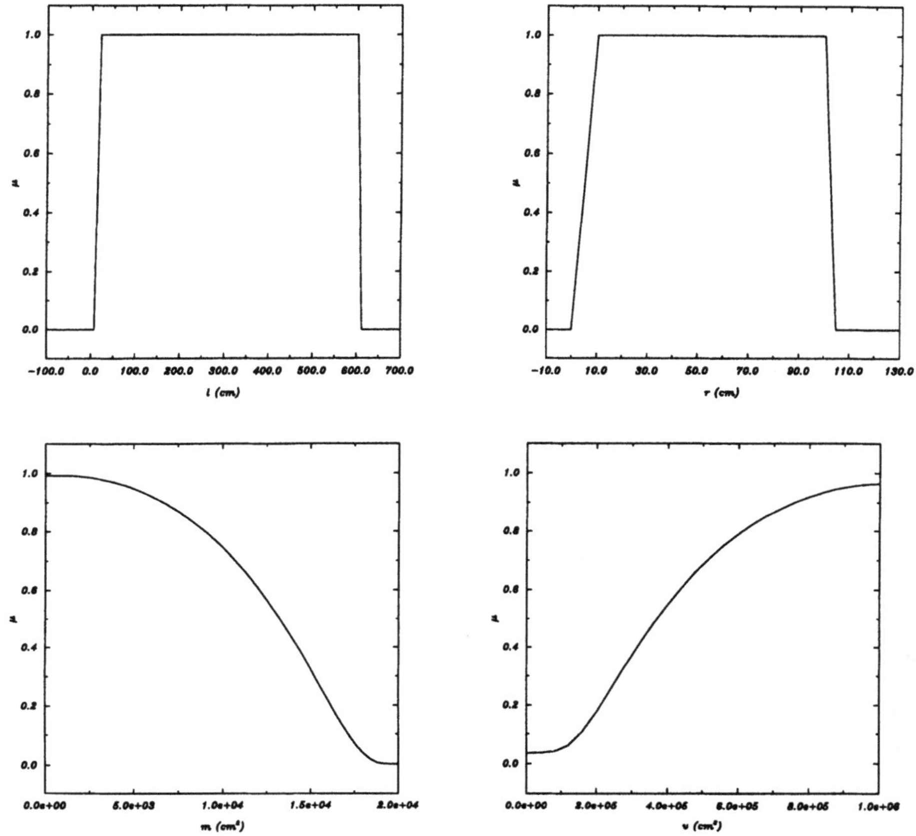

Figure 2 Preferences: Length l, Radius r, Metal volume m, Capacity v.

of functions, differential equations, integrations, computer programs, physical experiments, or subjective questioning.

As mentioned previously, stochastic uncertainty (noise) also is typically present in engineering design, and can be characterized by noise variables. A noise variable n_k might be the possible positioning of an operator switch, and so the alternatives may be finite. Alternatively, n_k might be a value of a manufacturing error on a design variable, and so the Noise Variable Space (NVS) may have a continuum of possibilities. In this case, n_k might be thought of as a vector within an NVS $\simeq \mathbb{R}^q$ represented with coordinates n^ℓ in some basis, for example. The approach described here can also include these effects, and

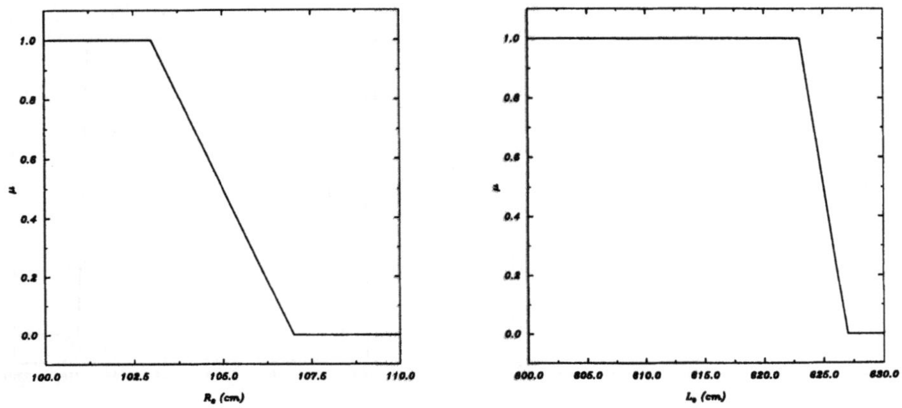

Figure 3 Preferences: Outer Radius R_O, Outer Length L_O.

can show how the level of noise present in a design affects the customers' and designers' preferences, and the trade-offs made in design decision-making.

In the tanks example, the following are noise variables, and will be discussed further in Section 4:

- material maximum stress (S),
- manufacturing variations in dimensions l and r (δl, δr),
- uncertainty in the welded joint efficiency (E),
- unknown applied pressure (P).

The next section briefly discusses the approach used in the M_oI for propagating imprecise information from the DVS to the PVS and Section 3 illustrates trade-offs (aggregations) that can be made between design and performance variables, all without considering noise. Noise and its consequences for engineering decision-making are presented in Section 4.

2 ENGINEERING DESIGN WITH IMPRECISION

After specifying design preferences $\mu(\vec{d})$ on the DVS and performance specifications (customer preferences) $\mu(\vec{p})$ on the PVS, the next step is to determine the induced values of $\mu(\vec{d})$ on the PVS (design preferences mapped onto the performance space), given by the extension principle [63]. This calculation is performed over the *entire set* of imprecise values specified, allowing the designer to more readily evaluate *sets* of designs.

$$\nu(\vec{p}) = \begin{cases} \sup\left\{\mathcal{P}(\mu_1(\vec{d}), \ldots, \mu_N(\vec{d})) \mid \vec{p} = f(\vec{d})\right\} \\ 0 \quad \text{if } \left\{\vec{d} \mid \vec{p} = f(\vec{d})\right\} = \emptyset \end{cases} \tag{1.10}$$

where sup over the null set is defined to be zero. (Aggregation operators ($\mathcal{P}$) will be discussed in the next section.) $\nu(\vec{p})$ is the combined design preference induced onto the PVS by use of the extension principle. The process of mapping the design preference $\mu(\vec{d})$ from the DVS to $\nu(\vec{p})$ on the PVS is referred to as *inducing* the design preference onto the PVS. Thus $\nu(\vec{p})$ is referred to as the *induced* design preference.

To calculate $\nu(\vec{p})$, the MoI uses an algorithm to compute Zadeh's extension principle. $\nu(\vec{p})$ is calculated by use of the *Level Interval Algorithm* (LIA), first proposed in [11] as the "Fuzzy Weighted Average" algorithm and also called the "Vertex Method", and extended by the authors in [38, 60]. The LIA uses α–cuts D_{α_k} in the DVS to calculate individual induced α–cuts P_{α_k} in the PVS:

$$\begin{aligned} D_{\alpha_k} &= \{\vec{d} \in \mathcal{X} \mid \mu_d(\vec{d}) \geq \alpha_k\} \\ &= [d_{1\,\min}^{\alpha_k}, d_{1\,\max}^{\alpha_k}] \times \ldots \times [d_{n\,\min}^{\alpha_k}, d_{n\,\max}^{\alpha_k}] \end{aligned} \tag{1.11}$$

$$\begin{aligned} P_{\alpha_k} &= \{\vec{p} \in \mathcal{Y} \mid \mu_d(\vec{p}) \geq \alpha_k\} \\ &= [p_{1\,\min}^{\alpha_k}, p_{1\,\max}^{\alpha_k}] \times \ldots \times [p_{q\,\min}^{\alpha_k}, p_{q\,\max}^{\alpha_k}] \end{aligned} \tag{1.12}$$

where $k = 1, \ldots, M$.

After having made the performance variable induced preference calculations, the designer can observe the imprecise performance achievable and proceed to evaluate the candidate designs. The design preferences induced onto the performance space:

$$\nu_{(l,r)}(m, v) = \min\left(\mu(l), \mu(r)\right) \tag{1.13}$$

are illustrated for the air tank design example in Figure 4 plotted on (m, v).

If the α–cut information is preserved in the LIA calculations, imprecise information can be propagated in both directions: "forward" from the DVS to the PVS; then subsequently "backwards" from the PVS to the DVS [39, 56]. This is "backwards" propagation of performance specification information and is shown for the tank example in Figures 5 and 6, as discussed further in Section 3.4.

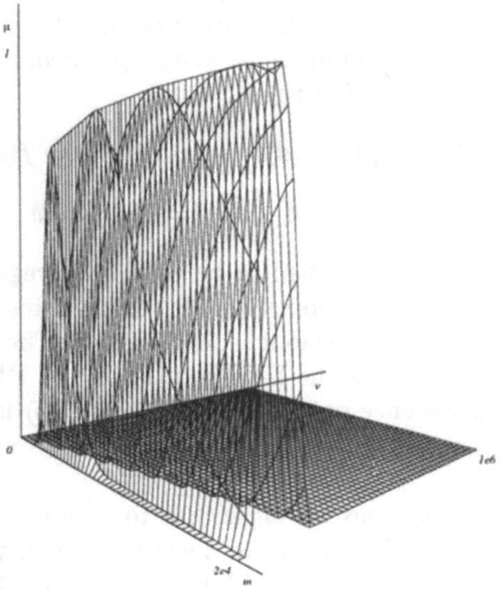

Figure 4 Preference for l and r induced onto the PVS (m, v).

3 COMBINING IMPRECISION IN DESIGN

One of the most challenging aspects of design decision-making is to compare different design alternatives. This is especially difficult because each alternative typically has many different attributes (such as weight, size, cost, etc.) that are not immediately comparable. Aggregating these incommensurate attributes to arrive at a rating for each design (and trading-off attributes with an aggregation method to arrive at the best rating for each alternative) is difficult. The traditional approach is to use a weighted sum of attribute values [1, 2, 4, 6, 10, 16, 20, 23, 31, 41, 42, 45, 46, 61]. This approach, however, has difficulties, as will be shown below.

With the Method of Imprecision, incommensurate attributes can be aggregated by combining the unrelated preference specifications: namely by combining the μ_k. An *overall preference* is a map $\mathcal{P}$:

$$\mathcal{P} : [0,1]^N \to [0,1] \tag{1.14}$$

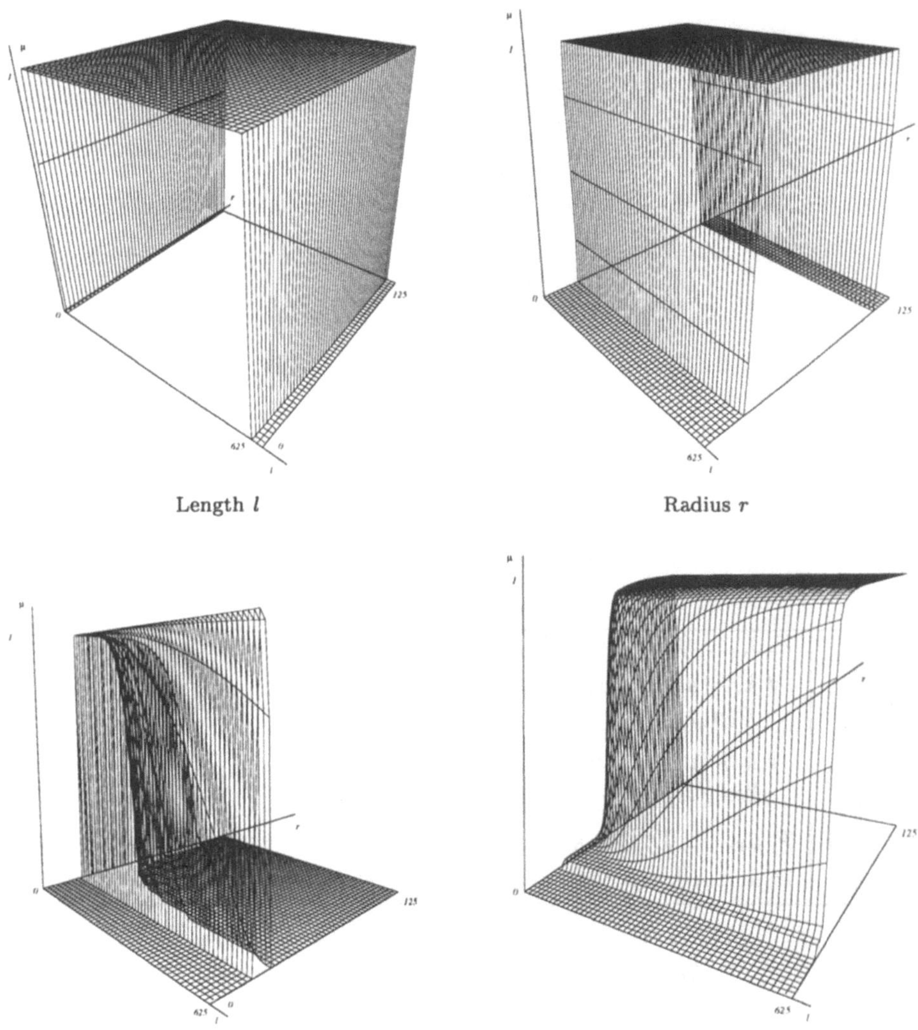

Length l Radius r

Metal Volume m Capacity v

Figure 5 Preferences on the DVS: l, r, m, v.

which reflects a designer's degree of overall satisfaction for a design. Using this map, the configuration among the options that best satisfies the designer (here denoted as $\vec{d}^*$) is:

$$\vec{d}^* : \mu(\vec{d}^*) = \sup\{\mathcal{P}(\mu_1, \ldots, \mu_N) \mid d \in \text{DVS}\}. \tag{1.15}$$

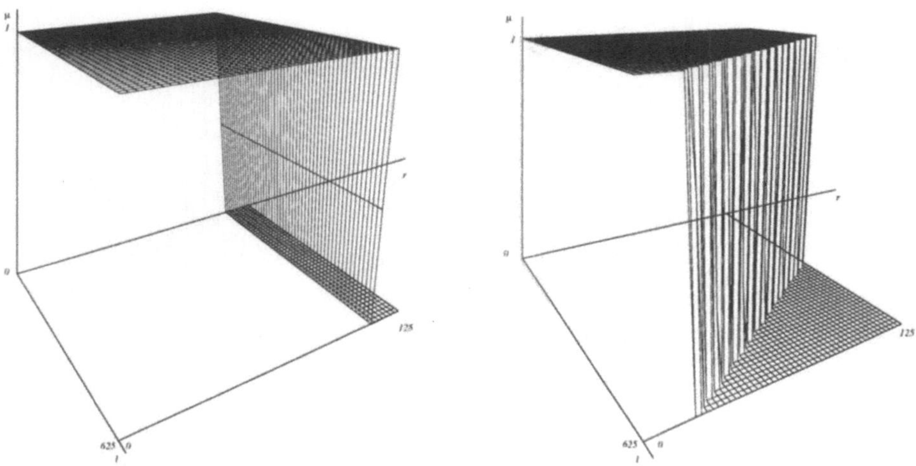

Outer Radius R_0 Outer Length L_0

Figure 6 Preferences on the DVS: R_0, L_0.

1	$\mathcal{P}(0,\ldots,0) = 0 \quad \mathcal{P}(1,\ldots,1) = 1$	(boundary conditions)
2	$\forall\, k, \mathcal{P}(\mu_1,\ldots,\mu_k,\ldots,\mu_N) \leq \mathcal{P}(\mu_1,\ldots,\mu'_k,\ldots,\mu_N) \text{iff} \mu_k \leq \mu'_k$	(monotonicity)
3	$\forall\, k, \mathcal{P}(\mu_1,\ldots,\mu_k,\ldots,\mu_N) = \lim_{\mu'_k \to \mu_k} \mathcal{P}(\mu_1,\ldots,\mu'_k,\ldots,\mu_N)$	(continuity)
4	$\mathcal{P}(\mu_1,\ldots,0,\ldots,\mu_N) = 0$	(annihilation)
5	$\mathcal{P}(\mu,\ldots,\mu) = \mu$	(idempotency)

Table 1 Overall Preference Resolution Axioms.

The remaining question is: Which aggregation functions ($\mathcal{P}$) are appropriate for engineering design? The authors have previously proposed a set of restrictions, based on product development criteria, that all aggregation functions must be consistent with for engineering design purposes [35]. These are shown in Table 1. This set of restrictions confines $\mathcal{P}$ to be a mixed connective. This means any decision made by the designer with $\mathcal{P}$ will result in ratings which are bounded by the worst (min) and best (max) ratings within a design. Table 1 is a list of necessary requirements and therefore defines many metrics, more are discussed in [35]. Two in particular are important for design: reflecting whether

6	$\mathcal{P}(a,b) = \mathcal{P}(b,a)$	(commutativity)
7	$\mathcal{P}(\mathcal{P}(a,b),c) = \mathcal{P}(a,\mathcal{P}(b,c))$	(associativity)
8	$\mathcal{P}(a,1) = a$	(identity)
9	$\mathcal{P}(a,b) \leq \min\{a,b\}$	(upper limit)

Table 2 *Min* Preference Resolution Axioms.

a designer wishes to make design decisions based on the worst case, or to make design decision with compensation among the goals. A family of aggregation functions to perform these trade-offs is presented in [43], including importance weightings.

3.1 Non-Compensating Trade-Offs

Suppose a designer wishes to trade-off to improve the lower performing attributes when selecting a configuration $\vec{d}^*$. Then to improve a design, there must be an increase in the preference level of the attribute whose preference is lowest. Thus the most appropriate method for combining multiple preferences in this way is $\mathcal{P} = min$.

$$\mu(\vec{d}) = \min\{\mu_1, \ldots, \mu_N\} \tag{1.16}$$

In this case, there are additional assumptions which must be made to restrict $\mathcal{P}$ to *min*, which a designer must consider. These additional restrictions have been developed elsewhere [14, 35]. Attributes aggregated with $\mathcal{P} = min$ are said to be combined with a non-compensating trade-off. The term "non-compensating trade-off" derives from the fact that overall preference for the design is dictated by the attribute with the lowest preference. No matter how high the preferences for the other attributes, they cannot compensate for the lowest one. The additional restrictions are shown in Table 2, for a two goal design. It has been shown that restrictions 1-3 and 5-8 in Table 1 and 2 are necessary and sufficient to restrict $\mathcal{P}$ to *min* [14]. Annihilation (restriction 4) can then be derived. It has been further shown that $\mathcal{P}$ is also restricted to *min* with restriction 7 (associativity) replaced by 9 (upper limit) [5]. Thus, in addition to the basic restrictions presented in Table 1, if one additionally assumes $\mathcal{P}$ is commutative and obeys the identity restriction, $\mathcal{P}$ must be *min*. Alternatively, associativity can be replaced by the upper limit restriction.

These additional restrictions must be justified for use in engineering design trade-offs. When a designer wants to improve the weakest link in a design, commutativity is a natural restriction. The order in which a designer combines objectives should not matter. For the same reason, $\mathcal{P}$ should be associative. If a designer combines sub-results which are then combined, the order in which this is done should not matter. Finally, the identity restriction is justified by considering the special case of when the other attribute (not the lowest performing) has a preference of 1.

3.2 Compensating Trade-Offs

The non-compensating trade-off strategy (as described above) is not always appropriate. If the preference for a design is drastically hindered by one variable and shifting the design-point to slightly reduce its preference greatly increases the others' preference, the modified design may be considered to produce a higher "overall" performance, even though the lower performing attribute is slightly reduced further. In some circumstances higher preferences for some attributes may legitimately compensate for lower preferences in other attributes. This can be accomplished consistently with Table 1 through the use of a product [62]:

$$\mu(\vec{d}) = \left(\prod_{k=1}^{N} \mu_k \right)^{\frac{1}{N}} \tag{1.17}$$

Higher performing attributes are allowed to compensate for lower performing attributes (in terms of preference) by (1.17). Again, additional restrictions must be accepted for a designer to use this combination, as developed in [35]. Attributes aggregated with $\mathcal{P}$ = ***product*** are said to be combined with a compensating trade-off. The term "compensating trade-off" derives from the fact that overall preference for the design is determined by attributes with higher preference compensating for attributes with lower preference. This trade-off strategy can be particularly appropriate in situations where a small decrease in the preference of the lowest preference attribute will greatly increase the preference for other attributes. This is also only appropriate where decreasing the lowest preference does not violate a constraint. These added restrictions are shown in Table 3.

Additional assumptions necessary to restrict $\mathcal{P}$ to the ***product*** have been considered previously [32, 34], and are shown in Table 3. These additional restrictions must also be justified for use in engineering design trade-offs.

6	$\mathcal{P}(a,b) = \mathcal{P}(b,a)$	(commutativity)
7	$\mathcal{P}(\mathcal{P}(a,b),\mathcal{P}(c,d)) = \mathcal{P}(\mathcal{P}(a,c),\mathcal{P}(b,d))$	(bi-symmetry)
8	$\mathcal{P}(\mu,a) < \mathcal{P}(\mu',a)$ iff $\mu < \mu'$	(strictness)

Table 3 *Product* Preference Resolution Axioms.

The first, commutativity, is a natural restriction: the order in which a designer combines objectives should not matter. For the same reason, $\mathcal{P}$ should be bi-symmetric. It should not necessarily be associative. If a designer combines sub-results which are then combined, the order in which this is done should not matter. But in a compensating strategy, the result of combining sub-results together carries more weight than an individual attribute (when the attributes are allowed to compensate in preference). A subset of attributes has more capacity to compensate than an individual attribute. Thus, associativity is an excessive restriction. Rather, equal numbers of attributes within subsets should carry equal weight (as a subset). Bi-symmetry restricts in this manner (for subsets of order 2).

With a compensating strategy, high preference on some attributes are to compensate for low preference on others. For example, higher preference on the cost of a design (i.e., it is cheaper than a different design) can compensate for any decreased preference for performance (the design does not perform as well as a different design). This implies any increase or decrease in preference of any of the attributes will change the compensation. Thus, the product combination must be strict.

These restrictions should be accepted if the designer desires to apply the compensating design trade-off strategy to an equally weighted design problem.

Thus there are two different strategies and formalizations for considering multiple goals in a design. These can be extended to considering different importances for each goal, for example, or to combinations of sub-goals [29, 43]. In general, however, the combination must remain consistent with the restrictions in Table 1.

3.3 Trade-Off Examples

Consider again the air tank design problem. The design space is formed by the two variables that the designer is free to choose, l and r. The problem is to find the values for l and r that maximize overall preference. For comparison, both a non-compensating and a compensating trade-off strategy will be presented and contrasted below. Both consider all goals to be equally important.

For the non-compensating trade-off strategy, l^* and r^* are to be found, where

$$\mu_c(l^*, r^*) = \sup\left\{\min\{\mu_l, \mu_r, \mu_{v(l,r)}, \mu_{m(l,r)}, \mu_{L_0(l,r)}, \mu_{R_0(l,r)}\} \mid (l, r) \in \mathbb{R}^2\right\}. \tag{1.18}$$

This will find the l^* and r^* by trading off the attributes to improve the lowest performing attribute (in terms of preference), even though the design variables and performance variables are incommensurate with each other.

For the compensating trade-off strategy, the problem to be solved is to find l^* and r^* where

$$\mu_c(l^*, r^*) = \sup\left\{\left(\mu_l \cdot \mu_r \cdot \mu_{v(l,r)} \cdot \mu_{m(l,r)} \cdot \mu_{L_0(l,r)} \cdot \mu_{R_0(l,r)}\right)^{1/6} \,\middle|\, (l, r) \in \mathbb{R}^2\right\}. \tag{1.19}$$

This will find the l^* and r^* by trading off the attributes cooperatively among each other, allowing the higher performing attributes to compensate for the lower performing attributes (in terms of preference), even though the design variables and performance variables are incommensurate with each other.

The preference combination results are shown in Figure 7. For the non-compensating trade-off strategy, the *min* of each individual preference across the design space is the resulting surface. This is shown at the top of Figure 7. The surface's maximum value in μ is the solution point to use (the most preferred l and r). For the compensating trade-off strategy, the individual preference surfaces are multiplied together as a product of powers for all points on the l, r plane. This is shown at the bottom of Figure 7.

As can been seen in the figures, the compensating strategy will produce higher overall preference than the non-compensating strategy. The two strategies will result in different design variable solution values: different l^*,r^* have the highest μ on the overall preference surfaces for the two different trade-offs applied to the hemispherical head design. The results are similar for the two different trade-offs applied to the flat head design. The non-compensating trade-off strategy sacrifices cost (m) preference to ensure that the capacity (v) specification is

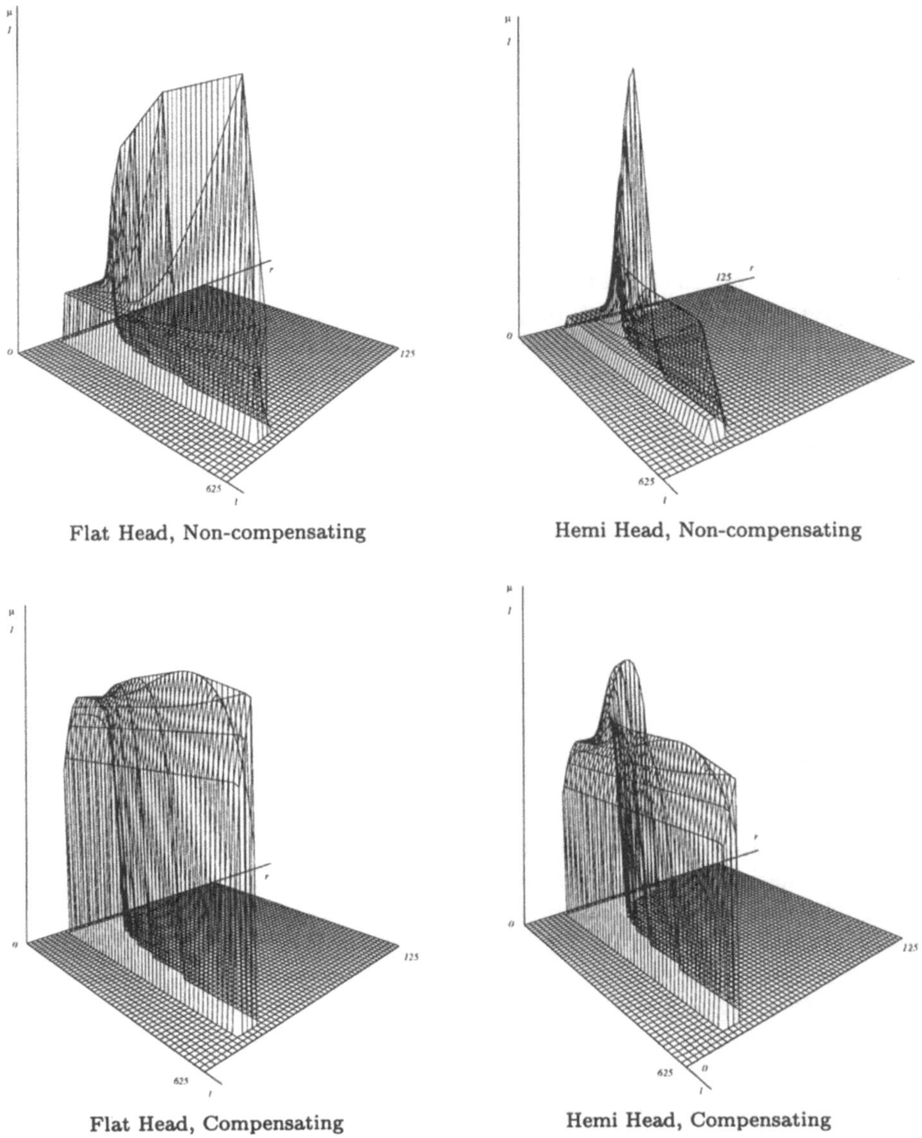

Figure 7 Flat and Hemi Head Tank Trade-Off Strategy Results.

met. Designing with a compensating strategy does the reverse: reduces cost (m) preference at expense of the capacity (v).

3.4 Optimization versus Iteration in Design

Suppose the designer wishes to combine all design and performance preferences into an overall measure to find a solution as discussed above. He could simply form an optimization of preference to search for the most preferred design configuration, as reflected by (1.18) or (1.19). This is rather artificial for design purposes, however, in that a designer is not sure of the preference functions, and would prefer to induce them through equations as advocated here. It would be of benefit if a designer were allowed to do the imprecise calculations and then make use of these to determine the final most preferred point in the DVS.

It has been shown earlier [39] that points which maximize overall preference can also be determined by another method. One can induce the preferences specified on the DVS $\mu(\vec{d})$ onto the PVS by the imprecision transformation to create $\nu(\vec{p})$, in a *forward* manner. There one can find the $\vec{p} \in$ PVS that maximize an overall preference, and then back map to the DVS simply by looking up the values of $\vec{d}$ used in the original forward mapping ν at the optimal $\vec{p}$. This is true even though the induced preference ν on the PVS involves only the preference $\mu(\vec{d})$, and does not consider any of the dependent set preferences $\mu(\vec{p})$. The results, however, are the points which maximize the overall preference μ_c: the supremum of the combination of the preferences specified both on DVS and PVS.

Thus the solution of (1.18) and (1.19), shown in Figure 7 on the DVS, can be obtained through a different iterative means using the Method of Imprecision. The design variable preferences are induced onto the PVS as shown in Figure 4. There they are compared with the performance specifications (customer preferences) shown in Figure 2. A point in the PVS which maximizes all the preferences is selected. The values of l and r which map to this point in the PVS using μ are the optimal values (l^*, r^*), just as if (1.18) or (1.19) had been used. This technique has significant advantages, because it presents visual information to the designer about the model. An optimization routine will produce a (hopefully globally optimal) solution, but it presents little information about the design and performance space, except at the optimal point in the DVS.

4 UNCONTROLLABLE VARIABLES IN DESIGN (NOISE)

This section now considers the influence of uncontrollable effects on the design. Design variables represent values that the designer is free to choose to best satisfy the performance requirements of the design, such as dimensions and material properties. Noise variables are uncontrollable effects that a designer does not have direct choice over, such as dimensional variations during manufacture, material property variations, etc. Since the values cannot be chosen by the designer, and instead are dictated by uncontrolled random or operational effects, they must all be included in the design calculations carefully so that the effect of noise is properly included in design trade-offs and decisions.

The Noise Variable Space (NVS) is defined to be an *measure space* $(NVS, \mathcal{B}, g)$, consisting of a set NVS of elements $\vec{n}$, a σ-algebra $\mathcal{B}$ of sets over the NVS, and an uncertainty measure $g : \mathcal{B} \to [0, 1]$.

An uncertainty measure is defined differently from a Lebesgue measure [22] or a fuzzy measure [13, 15, 47, 48] to take into account all possible ways a designer may wish to account for uncontrollable effects. The general restrictions to which any design method should conform when considering uncontrollable variables have been developed previously by the authors [37] and will be presented below. However, three specific measures stand out as important: the probability measure Pr, the possibility measure Π, and a necessity measure $\mathfrak{N}_\alpha$ (which is not the same as the necessity measure of Dubois and Prade [13]).

For clarity in this development, events in the NVS are assumed to have the structure of a σ-algebra. A σ-algebra $\mathcal{B}$ is naturally defined by the number of subsets within a designer's ability to characterize the NVS. While others [47, 48] have developed measures for sets without this structure, it suffices for design, and is historically well developed [22] for probability.

4.1 Measuring the Effects of Noise

An *integral* of the performance across a noise variable space (NVS) with respect to a measure of noise should be used. An integral of a function f :NVS $\to [0, 1]$ with respect to an uncertainty measure $g : \mathcal{B} \to [0, 1]$ over a σ-algebra of sets $\mathcal{B}$,

$$\int_N f(n)\, dg : (\text{NVS}, \mathcal{B}, g) \to [0, 1], \tag{1.20}$$

1	$\int_{NVS} \chi_N(n)\, dg = g(N)$	(identity)
2	$\int_{NVS} \alpha\, dg = \alpha$	(linear)
3	$f_1(n) \geq f_2(n) \Rightarrow \int_N f_1(n)\, dg \geq \int_N f_2(n)\, dg$	(increasing in f)
4	$N_j \supseteq N_k \Rightarrow \int_{N_j} f(n)\, dg \geq \int_{N_k} f(n)\, dg$ or $N_j \supseteq N_k \Rightarrow \int_{N_j} f(n)\, dg \leq \int_{N_k} f(n)\, dg$.	(monotone in N)

Table 4 Uncertainty Integral Restrictions.

is a map used to determine the performance rank of the function f over a subset N of the NVS as measured by the uncertainty measure g. This definition is quite general. For engineering design, there are specific restrictions the operation must satisfy, shown in Table 4, as developed in [37]. Any operation on an NVS that satisfies these conditions will be called an ***uncertainty integral***.

As with the use of axiomatic restrictions on the aggregation functions discussed earlier, the restrictions on uncertainty integrals for use in engineering design decision-making must also be justified. The first restriction in Table 4 requires that the uncertain integral of a characteristic function must produce the uncertain measure of the set.

The second restriction requires that the uncertain integral of a constant function must produce the constant. If the same performance results no matter what element in the NVS occurs, this level of performance should be the performance rating.

The third condition states the integral must be monotone in the argument. Higher performance at each element within the NVS results in a higher performance rating.

Finally, the last condition states that the uncertain integral must be monotone in inclusion for all N_j, N_k. As more of the NVS is considered, there is more ability for the noise to affect the performance rating. This differentiates the development from fuzzy integrals [47, 48], in that here the monotonicity is allowed in either direction, which proves necessary for design purposes.

4.2 Design Calculations with Noise

The performance to be measured across the NVS in the context of selecting the most preferred design variables is the overall preference $\mu(\vec{d}, \vec{n}) = \mathcal{P}(\mu_1, \ldots, \mu_N)$. Thus, an element $\vec{d} \in$ DVS can be rated using an operation satisfying the uncertainty integral restrictions:

$$\mu(\vec{d}) = \int_{NVS} \mu(\vec{d}, \vec{n}) \; dg \tag{1.21}$$

$\mu(\vec{d})$ as above is therefore the *expected* preference given the NVS. Further, the $\vec{n}$ are those used given a fixed $\vec{d}$. The NVS can generally vary with $\vec{d}$.

The most preferred points $\{\vec{d}^*\} \subseteq$ DVS are defined by

$$\vec{d}^* : \mu(\vec{d}^*) = \sup \left\{ \int_{NVS} \mathcal{P}(\mu_1, \ldots, \mu_N) \; dg \;\middle|\; \vec{n} \in \text{NVS},\; \vec{d} \in \text{DVS} \right\} \tag{1.22}$$

Thus, $\mu(\vec{d}^*)$ is the maximum *expected* preference, using the appropriate uncertainty integral. Again, generally the NVS can vary with $\vec{d}$. The NVS, $\mathcal{B}$, and g are all dependent on $\vec{d}$.

4.3 Probabilistic Noise

A particular uncertainty form can be used when the events are random. For example, inaccuracies in measurement and fabrication are usually modeled as random with a probability measure Pr. Such inaccuracies form what is now termed a ***probability space***. The algebraic structure of a probability space is not new [28, 50].

With a probability measure Pr [8, 12, 13, 28, 51], for events N_j, $N_k \in \mathcal{B}$ such that $N_j \cap N_k = \emptyset$, g is restricted to obey:

$$\begin{aligned} g(N_j) + g(\text{NVS} \backslash N_j) &= 1 \\ g(N_j) + g(N_k) &= g(N_j \cup N_k) \end{aligned} \tag{1.23}$$

Any uncertainty measure g satisfying (1.23) will be called a ***probability measure***, and denoted Pr.

As an example, consider again the design of a cylindrical tank. It may be known from past production the capability of the manufacturing process to

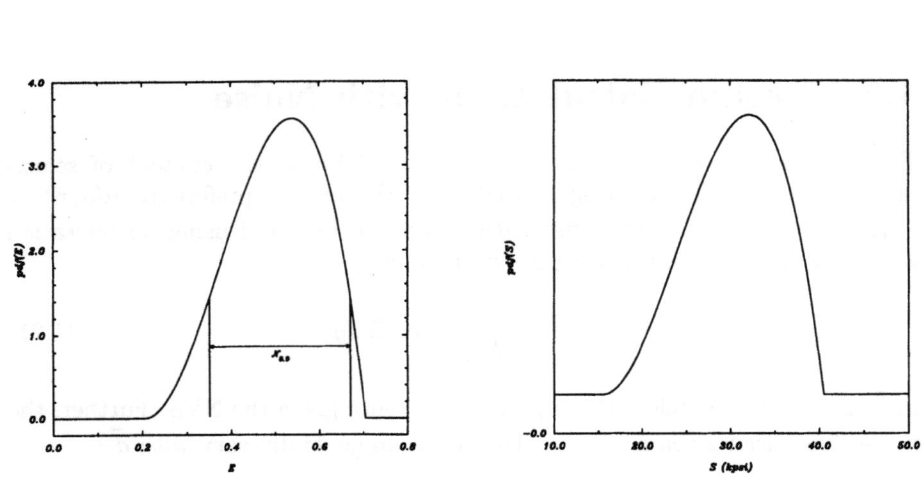

Measured Joint Efficiency (E) Material Maximum Stress (S)

Figure 8 Design Variable Probability Distributions.

weld the tank seams. This can be measured and represented with a probability distribution over a real variable representing joint efficiency E, as shown in Figure 8. Here, the probability measure is defined over the half open intervals, and is equal to the integrated probability density function over that domain.

Another source of uncertainty may arise from the material supplied. The material supplier will only guarantee the material properties to within a specified tolerance. For example, the maximum allowed stress can be measured on a representative sample of parts. This is another uncertainty that can be measured and represented with a probability distribution over a real variable representing maximum allowed material stress S, as shown in Figure 8.

The overall expected preferential performance at any design configuration becomes the integral of performance across the probability space.

$$\mu(\vec{d}) = \int_{NVS} \mathcal{P}(\mu_1, \ldots, \mu_N)\, dPr \tag{1.24}$$

This integral is the standard Lebesgue integral from measure theory [22]. Thus, $\mu(\vec{d})$ is the probabilistic expectation of $\mu(\vec{d}, \vec{n}) = \mathcal{P}(\mu_1, \ldots, \mu_N)$ across the probability space with respect to the probability measure Pr.

4.4 Possibility

Engineering design also encompasses decisions which are not made by a designer (and so are not controllable by the designer) and which are not random. For example, a post-manufacturing adjustment is often used to correct for uncontrolled manufacturing variations. A possibility measure Π can be used to represent such variables.

Possibilistic uncertainty can be modeled formally as a continuous *fuzzy measure*, as originally developed bySugeno Sugeno [47]. For possibilistic events $N_j, N_k \in \mathcal{B}$:

$$\begin{aligned} \max\{g(N_j), g(\text{NVS}\backslash N_j)\} &= 1 \\ g(N_j \cup N_k) &= \max\{g(N_j), g(N_k)\}. \end{aligned} \tag{1.25}$$

(1.25) defines a possibility measure [15].

As an example, consider again the design of a cylindrical tank. It may be known that the pressure applied to the tanks varies with customer application. From customer surveys, this may be understood in terms of belief of the customer. This can be represented as a possibility distribution over a real valued variable representing the applied pressure P as shown in Figure 9. Here, the possibility measure is defined over any interval, and is equal to the ***sup*** of the possibility density function over that domain.

These distributions are specified in a manner exactly the same as the preference functions were specified by the designer. The difference is that a customer, not the designer, supplies the preferences. Thus, the designer does not supply the preference uncertainty. For these designer-uncontrollable variables, the overall expected preferential performance is evaluated across a possibility space.

$$\mu(\vec{d}) = \sup\{\min\{\mathcal{P}(\mu_1, \ldots, \mu_N), \Pi(N)\} \mid N \in \{N_j\} \text{ disjoint} \subset \mathcal{B}\} \tag{1.26}$$

This integral is the Sugeno integral of possibility theory [47], expressed for the case where the NVS has the structure of a σ-algebra, as in engineering design. Note the ***sup*** is across the subsets of the partition $\{N_j\}$, and the limit as the partition becomes finer in $\mathcal{B}$ is used. The effects of possibility variables in engineering design were previously introduced and discussed by the authors in [36].

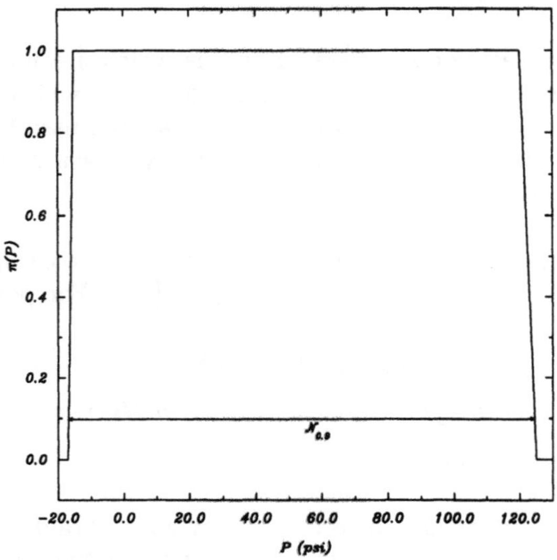

Figure 9 Applied Pressure Possibility.

4.5 Necessity

A final uncontrollable variable form represents uncertainty where proper performance must be ensured even under the worst case. For example, often the effects of a noise variable must be contained within 3 standard deviations of a probability distribution. This can be modeled with a *necessity measure*, different from that developed by Dubois and Prade Dubois and Prade [13]. Each necessary noise variable has a subset that a designer feels must be ensured: if the noise variable takes on a value outside the subset, then the design will not function properly. This subset $\mathcal{N} \in \mathcal{B}$ will be called the *necessary set*. Given this, any subset in $\mathcal{B}$ can be measured as necessary $\mathfrak{N}(N)$. If it lies within $\mathcal{N}$, the set is necessary, otherwise it is not. Thus, a *necessity measure* can be constructed given any necessary subset $\mathcal{N}_\alpha$ of the NVS. The measure can be defined by

$$\mathfrak{N}_\alpha(N) = \begin{cases} 0 & N \subseteq \mathcal{N}_\alpha \\ 1 & N \not\subseteq \mathcal{N}_\alpha. \end{cases} \tag{1.27}$$

Thus, only if all points within a set N are necessary does the set N become necessary, here necessary meaning $\mathfrak{N}_\alpha = 0$. The necessity measure therefore reflects the concern a designer has over the set being measured. Thus a low (zero) necessity reflects that a designer is concerned about the set; a high (one) necessity reflects that a designer is not concerned about the set.

This defines a crisp two-placed $\{0, 1\}$ ***necessity measure***. The actual uncertainty is incorporated into the extent of $\mathcal{N}$ in the NVS. Thus the uncertainty is incorporated into the specification of $\mathcal{N}_\alpha$ (the necessary set), not in $\mathfrak{N}_\alpha$ (the measure of a set in $\mathcal{B}$). α might be imprecise, for example, and the membership function for α included in $\mathcal{P}$.

A necessary measure is not a fuzzy measure, it fails to satisfy the fuzzy measure conditions. It is what will be termed a ***dual***-fuzzy measure. That is, the set inclusion property is reversed from a fuzzy measure g. $\mathfrak{N}_\alpha$ instead obeys:

$$\begin{aligned} &i) \quad \mathfrak{N}_\alpha(\text{NVS}) = 0, \qquad \mathfrak{N}_\alpha(\emptyset) = 1 \\ &ii) \quad N_i \subseteq N_j \Rightarrow \mathfrak{N}_\alpha(N_i) \geq \mathfrak{N}_\alpha(N_j), \qquad N_i, N_j \in \mathcal{B}. \end{aligned} \tag{1.28}$$

Clearly $\mathfrak{N}_\alpha$ is not necessarily continuous.

The problem is thus well formed, provided the set $\mathcal{N}_\alpha$ can be identified. How much of the NVS should be ensured? For an NVS $\simeq \mathbb{R}^q$, this has typically been done with *confidence intervals* [50]. This formalism, however, assumes that an underlying uncertainty measure (Pr or Π) is constructed from a density function (*pdf* or π), which requires an ordering on the NVS. In any case, the percentage of the uncertainty space to be satisfied is identified as the ***confidence factor***, and will be denoted α.

For example, in the case of NVS $\simeq \mathbb{R}$, and *pdf* as the normal distribution, $\mathcal{N}_\alpha$ becomes a class interval:

$$\mathcal{N}_\alpha = [E - r, E + r]$$

where E is the expected value, and r is a radial distance from the expected value such that $Pr([E - r, E + r]) = \alpha$. Further examples in the engineering design domain are demonstrated in [32, 37].

As an example, consider again the cylindrical tank problem. It may be decided that the error effects from the maximum allowed stress and the applied pressure must be ensured against to a confidence level of 90%. These two variables are then identified as necessary noise variables, and their distributions are examined and the set $\mathcal{N}_{0.9}$ identified. This is shown in Figures 8 and 9 for the maximum stress and applied pressure, respectively. Recall that the maximum stress is a probability random variable, and that the applied pressure is a possibility uncertainty. Necessity is a concept that can be applied to either type of uncertainty.

The overall expected preferential performance now becomes the integral of performance across the necessity space.

$$\mu(\vec{d}) = \inf\{\max\{\mathcal{P}(\mu_1, \ldots, \mu_N), \mathfrak{N}_\alpha(N)\} \mid N \in \{N_j\} \text{ disjoint} \in \mathcal{B}\} \quad (1.29)$$

This integral of performance provides the worst case performance across the necessity space, as measured by the necessity measure. It can be verified that this definition is indeed an uncertainty integral (defined in Table 4). Thus, $\mu(\vec{d})$ is the necessary expectation of $\mu(\vec{d}, \vec{n}) = \mathcal{P}(\mu_1, \ldots, \mu_N)$ across the necessary space with respect to the necessary measure $\mathfrak{N}_\alpha$.

Notice that this necessity measure is not the same as the necessity measure as developed by Dubois and Prade [13]. Their necessity measure reflects the state of "perhaps not possible" and so is essentially a restatement of the possibility measure: $Ne(N) = 1 - \Pi(N)$. Both can be exactly determined from the other. The necessity measure used in the MoI rather reflects the idea of "must be satisfied" which has nothing to do with being possible. Therefore, it should be no surprise the resulting formalisms are different.

5 HYBRID UNCERTAINTY

For problems with multiple uncertainty forms, the previous definitions must be combined. Such a combination is possible, but requires making explicit the *precedence relation* among the variables, to specify an evaluation order among the different formal operators.

For example, in the air tank design problem, there is imprecision from the designer making a selection of the values of design variables l and r. There is a random uncertainty in the joint efficiency E. There is a necessary random uncertainty to be ensured against in the maximum allowed stress S. There is a necessary possibility to be ensured against in the customer applied pressures P. There is a problem, however, in determining the order in which to combine the different integration operations for each of the different uncertainty forms.

To make this determination, the temporal order in which specific values are determined for the variables in a design, be they design variables, noise variables, or possibility variables, is used. This order is termed the *precedence relation* among the variables.

Typically, the precedence relation is:

- design variables are determined first,
- purchased material and manufacturing noise variables occur subsequently in the attempt to achieve the design variable targets,

and finally

- customer possibility variables are set according to the customer preferences.

In the air tank problem, the precedence relation is

$$(l,r) \prec S \prec (\delta l, \delta r) \prec E \prec P$$

That is, first the design variables (l,r) are specified. Then the material S is specified, then the parts are fabricated with errors $(\delta l, \delta r)$, then the parts are joined with error E, and finally the pressure is applied in use P.

With the precedence relation established, one can combine the previous definitions. Care must be taken to combine the equations in their proper order: according to the precedence relation.

For example, reconsider again the air tank design example. One can determine the overall expected preference across the design space, considering the uncertainty and the appropriate integration operations. The order to apply the integration operations is defined by the precedence relation.

$$\begin{aligned}\mu(l^*, r^*) \quad = \quad & \sup \{ \textstyle\int_S \int_{\delta l} \int_{\delta r} \\ & \inf \{ \mathcal{P}\{\mu_l, \mu_r, \mu_v, \mu_m, \mu_{L_0}, \mu_{R_0}\} \mid E, P \in \mathcal{N}_{0.9} \} \\ & \times pdf(\delta r)\ pdf(\delta l)\ pdf(S) \times d(\delta r)\ d(\delta l)\ dS \\ & \mid (l,r) \in \mathbb{R}^2 \}\end{aligned} \tag{1.30}$$

The results of this are shown in the bottom two graphs in Figure 10 for the non-compensating design trade-off strategy. These should be compared with the top two graphs in the same figure. The modulating effects of noise on the preferences functions can clearly be seen by this comparison.

The foregoing example illustrates the use of the Method of Imprecision, and the effects of design imprecision, trade-off strategies (aggregation), uncontrollable effects (noise) including probabilistic, possibilistic, and necessary effects. The final results of all of these imprecision and uncertainty effects on the pressurized air tank design problem are shown in Figure 10. Note that this figure shows the combined influence of all of the effects listed above, including noise, mapped back onto the design space (l,r). These results provide information to an

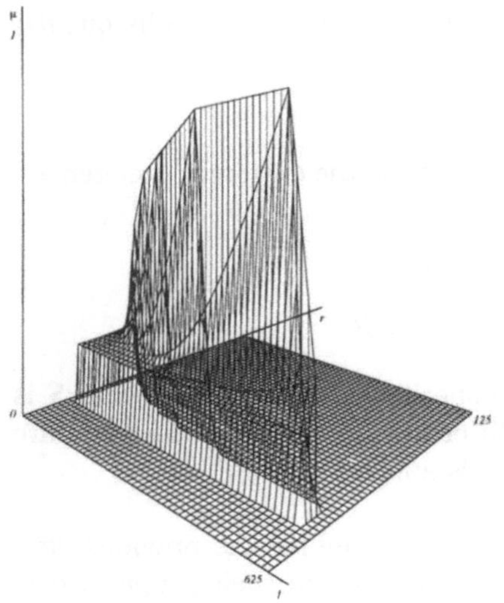

Flat Head, Non-compensating (no noise)

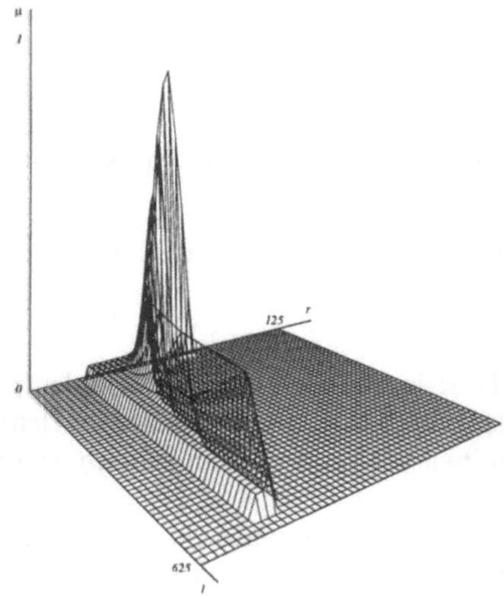

Hemi Head, Non-compensating (no noise)

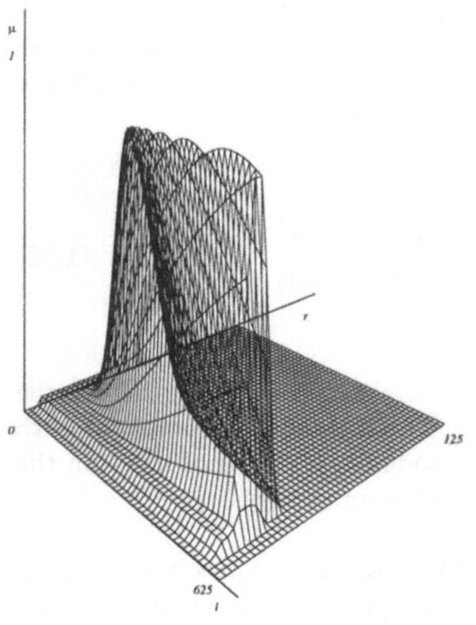

Flat Head, Non-compensating (with noise)

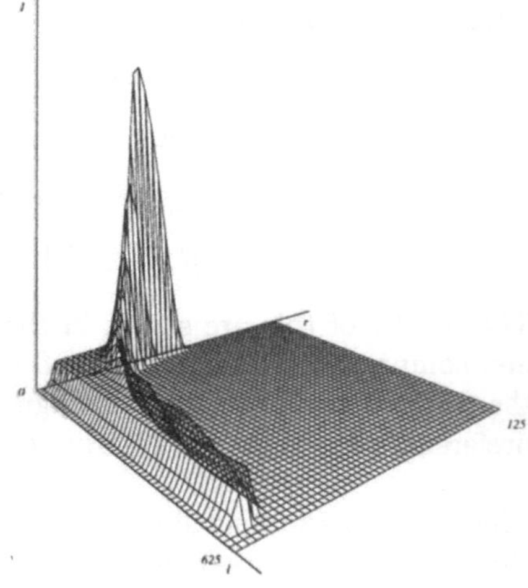

Hemi Head, Non-compensating (with noise)

Figure 10 Preferences for the Flat and Hemi Head Tanks.

engineering designer on a significant region of the design space, and indicate the range over which each alternative will operate satisfactorily. As shown in the figure, the hemispherical head design has a significantly smaller region in which it will operate satisfactorily than the flat head design. Furthermore, the combined peak preference for the two designs are very nearly at the same level. Thus the hemispherical head design appears to present no advantages over the flat head air tank design.

This interesting and valuable engineering design information is made available by the Method of Imprecision, without the need to perform a detailed design of each alternative. Thus, the M_oI meets its primary goal, of facilitating engineering design decision-making at the preliminary stages, when information is vague and imprecise.

The preceding air tank example is far simpler than most real engineering design problems. Its simplicity produces a valuable illustration, but misses many of the complexities of design in the real world. The example presented in the following section, the design of a braking system for a passenger vehicle, illustrates a design problem with many more complexities, and is therefore considerably more realistic.

6 A BRAKE DESIGN APPLICATION

In this section, a brake design is presented to illustrate the M_oI's iterative imprecision approach for a design problem with realistic design complexities, e.g., geometric and algebraic non-linearities in the performance variable expressions. In addition, simultaneous evaluation of multiple (alternative) configurations will be shown. These alternatives are evaluated using a non-compensating trade-off function. Noise, possibilistic uncertainty, and necessity are not explicitly included for the sake of brevity.

6.1 Design Problem Statement

Brakes perform the function of transforming the energy of a moving vehicle into heat (usually by use of friction) over an interval of time. Many factors must be considered when designing such devices. Of primary importance is the ability of the device to avoid destructive temperature rises while still dissipating wide ranges of output power.

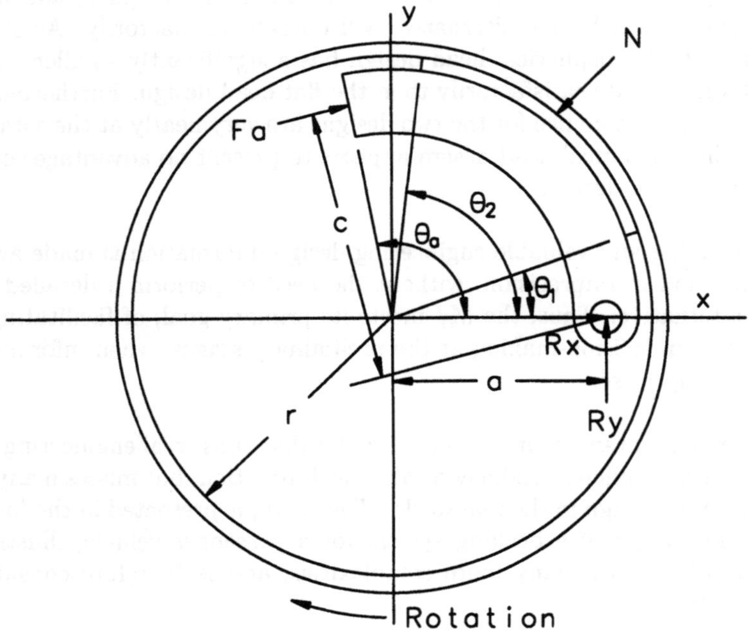

Figure 11 Drum Brake System.

The problem here is to design a braking system for a passenger vehicle that will adequately stop the vehicle for a certain range of speeds and which will not degrade appreciably (i.e., 'fade' due to temperature rise) over time. Two alternative design configurations for the problem (Figures 11 and 12) are considered.

The first, illustrated in Figure 11, is an internally actuated, pivoted rim brake (a drum brake), where only one shoe is shown in the figure. The other shoe would be placed in a symmetrical arrangement with the one shown. Formulating a mathematical representation of the design alternatives, four primary performance variables are chosen for the drum brake [44]: actuating force F_a, transmitted torque $\mathcal{T}$, temperature rise T_r and temperature decay profile (proportional to heat rejection) T_d. In equation form, this PVS may be expressed as:

$$F_a = \frac{M_N - M_\eta}{c}, \tag{1.31}$$

$$\mathcal{T} = \frac{\eta p_a b r^2}{\sin \theta_a}(\cos \theta_1 - \cos \theta_2), \tag{1.32}$$

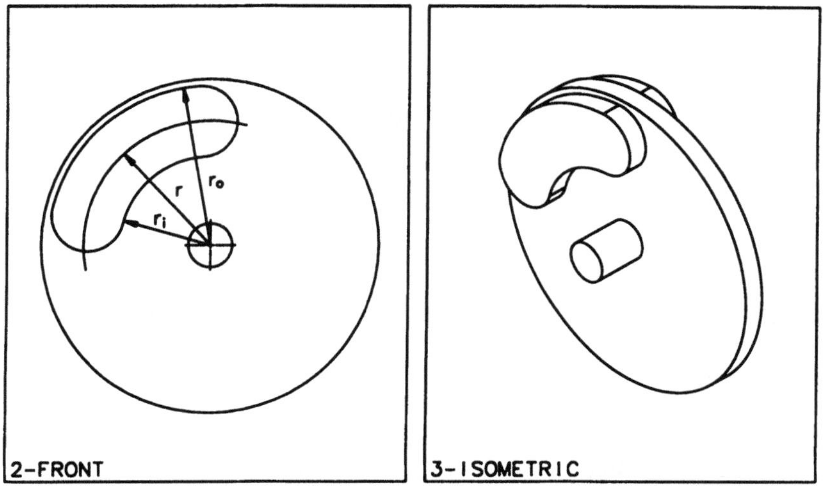

Figure 12 Disk Brake System.

$$T_r = \frac{E}{CM_b}, \tag{1.33}$$

$$T_d = T_r - \left(T_r e^{-\frac{2\pi L_w t}{\left[\left(\frac{\ln(R_1/r)}{k} + \frac{1}{rh}\right) M_b C\right]}} \right). \tag{1.34}$$

In order to completely describe the brake drum problem, the expressions for total system energy E, as well as A, B, M_N, M_η, I_d, and M_b must be given:

$$E = \frac{1}{2} I_d \omega^2 + \frac{1}{2} M_v V^2, \tag{1.35}$$

$$A = \frac{1}{2}(\sin^2\theta_2 - \sin^2\theta_1), \tag{1.36}$$

$$B = (\frac{\theta_2}{2} - \frac{\sin 2\theta_2}{4}) - (\frac{\theta_1}{2} - \frac{\sin 2\theta_1}{4}), \tag{1.37}$$

$$M_\eta = \frac{\eta p_a b r}{\sin\theta_a}[r(\cos\theta_1 - \cos\theta_2) - \frac{a}{2}(\sin^2\theta_2 - \sin^2\theta_1)], \tag{1.38}$$

$$M_N = \frac{p_a bra}{\sin\theta_a} B, \tag{1.39}$$

$$I_d = \frac{1}{2} M_b (R_1^2 + r^2), \tag{1.40}$$

$$M_b = \pi \rho L_w (R_1^2 - r^2). \tag{1.41}$$

The design variables for the problem are as follows: the pivot distance a; the radius r; the material angles θ_1 and θ_2; the coefficient of friction η; the material pressure p_a; the actuating force distance c; the face width b; the angular velocity ω; the specific heat C; the vehicle mass M_v; the outer drum radius R_1; the drum width L_w; the drum material conductivity k; and the coefficient of convection h. Note that the velocity of the vehicle V is not an independent design variable since it depends directly upon ω. Note further that the imprecision due to the "constant" ρ is viewed as negligible for this design. If desired, ρ may be included as a design variable (instead of a constant) to verify this assumption.

The second configuration to be considered is a disk brake system, illustrated in Figure 12, comprising two simultaneously actuated brake pads which close on both sides of the disk. The same *PVS* exists for the disk configuration as for the drum configuration. However, in this case, the actuating force and torque for the system depend upon an assumption: uniform wear or uniform pressure. Denoting the actuating force and torque for uniform wear as F_w and T_w, and similarly denoting these properties for uniform pressure as F_p and T_p, the governing performance variables may be expressed as:

$$F_w = 2\pi p_{max} r_i (r_o - r_i), \tag{1.42}$$

$$F_p = \pi p_p (r_o^2 - r_i^2), \tag{1.43}$$

$$\mathcal{T}_w = \pi \eta p_{max} r_i (r_o^2 - r_i^2), \tag{1.44}$$

$$\mathcal{T}_p = \frac{2}{3} \pi \eta p_p (r_o^3 - r_i^3), \tag{1.45}$$

$$T_d = T_r - \left(T_r e^{-\frac{h 2\pi r_o^2}{M_b C}} \right). \tag{1.46}$$

The energy expression and temperature rise performance variables are the same as the drum brake configuration, except the expression for the mass of the brake parts M_b differs:

$$M_b = \pi \rho r_o^2 L_{w1}. \tag{1.47}$$

The corresponding design variables for the disk brake problem are: maximum pressure for uniform wear p_{max}; outer radius of disk r_o; inner point of contact of brake shoe r_i; coefficient of friction η; pressure for uniform pressure p_p; specific heat C; convection coefficient h; width of disk L_{w1}; angular velocity ω; and vehicle mass M_v.

6.2 Performance Specifications

Given the four performance variables for each brake configuration as above, preliminary performance criteria may be specified. For example, the customer may associate a limiting value (performance specification or requirement) F^r for the actuating force performance variable F_a, F_w, or F_p. Such a limiting value represents a maximum actuating force that can be used in the design, i.e., F must be less than or equal to F^r where:

$$F \leq F^r = 25.0 \text{ kN}.$$

Consider the next measure of performance: torque ($\mathcal{T}$, $\mathcal{T}_w$, or $\mathcal{T}_p$). Instead of a maximum as with the actuating force, the torque must meet a minimum limit for the design. Specifically, the frictional torque developed by the brake system must be matched by the frictional torque developed between the road and the tires. A common value to use for the torque on the brake is three-fourths (3/4) of the average force (weight) on one of the wheels multiplied by the radius of the wheel. The resulting minimum performance torque for the brake design may be expressed as:

$$\mathcal{T} \geq \mathcal{T}^r = \tfrac{3}{4} F_{wheel} \times R_{wheel} \text{ kN-m},$$

where F_{wheel} equals the force on the wheel, R_{wheel} equals the wheel radius. Notice that a single numerical performance specification value does not result for $\mathcal{T}^r$. Because $\mathcal{T}^r$ depends on the force on one wheel which in turn depends on the mass of the vehicle M_v, the performance specification $\mathcal{T}^r$ is an induced imprecise function of the vehicle mass. (The radius of the wheel is assumed to be specified as a crisp constant for the design.) After the design variable M_v is determined, $\mathcal{T}^r$ may be calculated and updated if M_v subsequently changes.

A specification for the maximum temperature rise may be obtained from empirical data:

$$T_r \leq T_r^r = (260 - T_{ambient}) \quad {}^{\circ}\mathrm{C}.$$

The performance variable T_d does not have either a single performance specification value or an imprecise one. A qualitative specification (denoted by T_d^r) must therefore be used: ***maximize heat dissipation in order to minimize the possibility of destructive temperatures for the frictional material.***

These four performance specifications, cast as: single values; a preference function; and a qualitative statement, form the set of performance specifications for the PVS. This space is used to (1) rate each design configuration individually according to performance, (2) compare the alternatives in order to determine major differences, and (3) evaluate whether each configuration should be carried to the next stage in the design process.

6.3 Design Variable Space

Triangular preference functions are used in this illustration (for simplicity) for the imprecise design variables in both the drum brake and disk brake configurations, however, any function that represents preference can be used. Table 5 lists the necessary data for constructing the brake design's preference functions, where the three data values for each DV have the following meaning: left-extreme value for preference of zero (0), peak value for preference of one (1), and right-extreme value for preference of zero (0). $\tilde{x}$ will henceforth be used to denote a fuzzy number. Again, various interpretations exist by which a designer can assign the preference.

6.4 Induced Performance Variables

With the design variables specified, the induced performance variables may be determined using the performance variable equations listed above. The schematic representations of the induced preference functions for both the drum and disk brake alternatives are shown in Figures 13 through 16.

DVs (units)	$\mu = 0$	$\mu = 1$	$\mu = 0$
a (m)	0.010	0.020	0.030
b (m)	0.025	0.050	0.075
η	0.10	0.30	0.50
ω (rps)	0.0	15.0	30.0
p_a (kPa)	300.0	1100.0	1900.0
r (m)	0.10	0.15	0.20
R_1 (m)	0.005	0.015	0.030
θ_1 (deg.)	0.0	0.0	30.0
θ_2 (deg.)	90.0	120.0	150.0
k (W/m °C)	50.0	65.0	80.0
h (W/m^2 °C)	120.0	150.0	180.0
M_v (kg)	200.0	900.0	1600.0
L_w (m)	0.025	0.060	0.100
C (J/kg °C)	400.0	500.0	900.0
p_{max} (kPa)	300.0	1100.0	1900.0
p_p (kPa)	300.0	1100.0	1900.0
r_o (m)	0.10	0.15	0.20
r_i (m)	0.000	0.025	0.040
L_{w1} (m)	0.0125	0.0250	0.0375

Table 5 Brake Example: Imprecise Design Variable Data.

Drum Brake Induced Performance Variables

Using the induced performance variables for the drum brake configuration (Figure 13), the performance variables may be rated against the performance specifications, F^r, T^r, T_r^r, and T_d^r. The top left graph in Figure 13 represents the induced preference $\tilde{F}_a$ for the drum brake actuating force performance variable, from (1.31). The value of $\tilde{F}_a$ for preference of one ($\nu_{\tilde{F}_a} = 1$) is $F_a = 1.94$ kN. Comparing this result with F^r, $\tilde{F}_{a(at\nu_{\tilde{F}_a}=1)}$ is less than F^r by at least a factor of ten. The imprecision in the actuating force output (i.e., the change of $\tilde{F}_a$ with respect to preference $\nu_{\tilde{F}_a}$) is very small when compared to the difference of F^r and $\tilde{F}_{a(at\nu_{\tilde{F}_a}=1)}$. In fact, the actuating force for the drum brake does not exceed F^r until $\nu_{\tilde{F}_a} = 0.2$, implying that a change in any, some, or all of the design variables with preference greater than or equal to 0.2 for one side of the design variable preference function will still satisfy F^r. Thus, the drum brake configuration satisfies the actuating force performance specification for the most preferred values of the design variables, as well as design variables with preference far off the peaks.

The negative force results found for actuating force, $\tilde{F}_a$, in Figure 13 must also be considered. In this case, a geometrical dependency inherent in the problem, i.e., the relationship between the input variable a and the radius r, can produce negative actuating forces. Although such negative force values are perfectly correct physically, control problems arise of lifting the shoe from the drum with F_a instead of applying the shoe to the drum with this force. These performance results for $\tilde{F}_a$ show that care must be taking when using one or more design variables with preference less than 0.3.

The induced preference for the torque performance variable $\mathcal{T}$, from (1.32), is shown at the top right of Figure 13. Before this result may be compared with the performance specification, $\mathcal{T}^r$ must be calculated using the design variable M_v. Figure 14 shows the result. (The right side of the curve in Figure 14 corresponds to increasing values of M_v, while the left side corresponds to the converse.) The value of $\tilde{\mathcal{T}}^r$ for preference of one is $\mathcal{T}^r = 0.579$ kN-m. Comparing this value with the induced preference $\mathcal{T}$, $\tilde{\mathcal{T}}_{(at\nu_{\mathcal{T}}=1)}$ is less than $\mathcal{T}^r$; thus the torque results for the most preferred inputs *does not* satisfy the performance specification. Values to the right of the peak of $\mathcal{T}$ in Figure 13 are found to satisfy $\mathcal{T}^r$; in fact, for $\nu_{\mathcal{T}} = 0.9$, $\mathcal{T} = 0.736$ kN-m. The imprecise performance specification for $\mathcal{T}^r$ in Figure 14 shows that if M_v is the variable that changes, the corresponding $\mathcal{T}^r$ to the right of the peak is satisfied. The induced torque results for the drum brake configuration demonstrate that further analysis must be performed to satisfy the performance specifications. Specifically, for the most preferred design variable values, as well as values with preference off the peaks, the torque requirement may be barely satisfied or not at all.

The temperature rise induced set for the drum brake configuration may be found in Figure 13. Using a constant ambient temperature of $T_{ambient} = 24°C$, the corresponding performance specification for temperature rise is $T_r^r = 236°C$. As expected for any braking system, a complete stop from the design speed of $\omega = 15$ rps (or approximately 35 m/s) produces a temperature rise, $\tilde{T}_{r\,(at\,\nu_{\tilde{T}_r}=1)}$ $= 35°C$, which falls well within the performance specification T_r^r. It may be inferred that a few stops in succession at this speed will also result in an accumulative temperature rise which meets T_r^r. However, for many multiple braking actions in succession, a nominal value of $T_r = 35°C$ for each action will accumulate to temperatures approaching the performance requirement. Moreover, considering the induced values for $\tilde{T}_r$, a small change in $\nu_{\tilde{T}_r}$ on the right portion of the curve (say $\nu_{\tilde{T}_r} = 0.7$) results in temperature rises which quickly approach

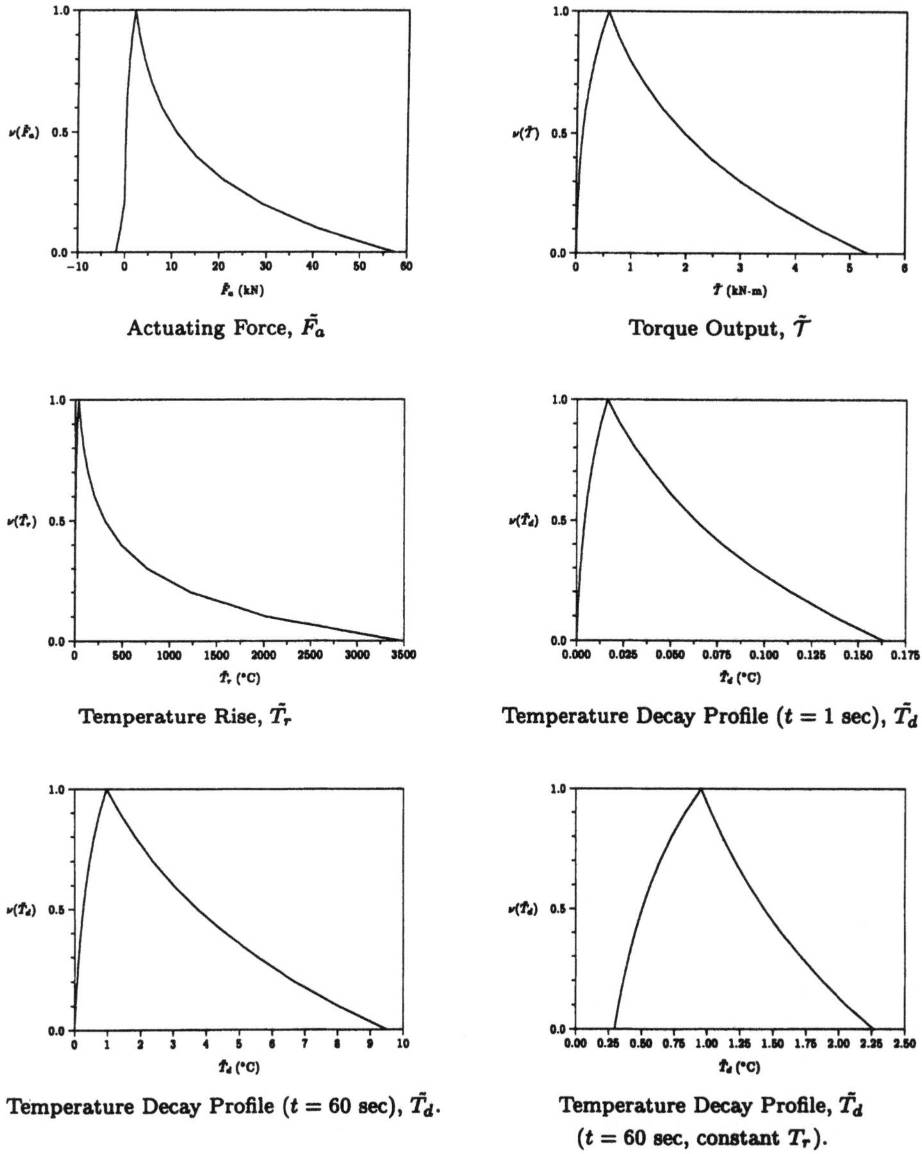

Actuating Force, $\tilde{F_a}$

Torque Output, $\tilde{T}$

Temperature Rise, $\tilde{T_r}$

Temperature Decay Profile ($t = 1$ sec), $\tilde{T_d}$

Temperature Decay Profile ($t = 60$ sec), $\tilde{T_d}$.

Temperature Decay Profile, $\tilde{T_d}$ ($t = 60$ sec, constant T_r).

Figure 13 Drum Brake: Induced Preferences.

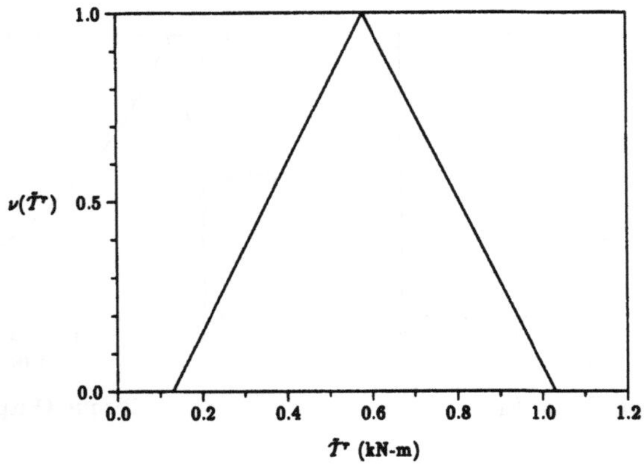

Figure 14 Preference: Torque Performance Specification, $\tilde{T}^r$.

the performance specification in a few stops. This means that the imprecision of $\tilde{T}_r$ is very dramatic for increasing values of T_r to the right of the peak. While all values to the left of the peak of $\tilde{T}_r$ satisfy T_r^r, the peak of $\tilde{T}_r$ as well as values to the right demonstrate that a small decrease in preference of any or all of the design variables may produce undesired design results. Temperature rise is therefore a critical performance variable, requiring special attention in later stages of the design with respect to the design variables.

Figure 13 contains the induced set for the temperature decay profile $\tilde{T}_d$ for a time of $t = 1$ sec and for a design preference $\tilde{T}_r$ with peak at $T_r = 60°C$ and right and left boundaries of $T_r = 175°C$ and $T_r = 0°C$, respectively. At $\nu_{\tilde{T}_d} = 1.0$, the induced performance variable value is $T_d = 0.016°C$. The bottom left graph in Figure 13 contains the output $\tilde{T}_d$ after $t = 60$ sec, where $\tilde{T}_{d\,(at\ \nu_{\tilde{T}_d}=1)} = 0.96°C$. These results demonstrate that the temperature decay is approximately linear for the time interval being considered (within 60 seconds). The resultant values for $\tilde{T}_d$ where $\nu_{\tilde{T}_d} = 1$ also show that the temperature decay is very slow for the most highly preferred choices of the design variables. Analyzing the temperature decay profiles in Figure 13, it appears that much greater (by an order of magnitude) temperature rejection may be obtained by choosing design variable values with small preference; however, these output values are not only dependent on the design variables under consideration (L_w, R_1, r, k, h, and C) but also upon the induced performance variable T_r. Only the design variables may be altered to obtain required performance, not performance such as T_r. Thus, to determine how the $\tilde{T}_d$ induced set varies for different design variables, T_r must be set to a representative crisp value (e.g., $T_r = 60°C$), and the induced set T_d must be recalculated using this value. The bottom right graph in Figure 13 shows the result of such a calculation for time $t = 60$ sec. Notice that when compared with the bottom left graph in Figure 13, greater temperature decay occurs as T_r increases, and that the imprecision due to T_r is the major contributor to the imprecision in $\tilde{T}_d$, implying that any change of any or all of the design variables will not significantly change the amount of heat dissipation.

Relating this discussion to T_d^r, the heat rejection of the drum brake does not compare well to the possible temperature rise values obtained in the system. Because of the small imprecision produced by changing one or many of the design variables, very little may be done to maximize the heat dissipation beyond that given by the peak of the $\tilde{T}_d$ induced curve. The γ*-level* measure (to be discussed below in Section 6.5) will provide further evaluation of the T_d^r performance specification.

Disk Brake Induced Performance Variables

The induced performance variableperformance variables results for the disk brake configuration may be found in Figures 15 and 16. This illustrative example will show the ease by which the different assumptions of uniform wear or uniform pressure may be simultaneously evaluated for the disk configuration. An alternative means of performing this evaluation would be to classify each assumption case as a separate configuration.

The actuating force variables, with given performance expressions (1.42) and (1.43), have induced preferences shown at the top of Figure 15. The peaks of the curves are: $\tilde{F}_{w\,(at\ \nu_{\tilde{F}_w}=1)} = 21.6$ kN and $\tilde{F}_{p\,(at\ \nu_{\tilde{F}_p}=1)} = 23.8$ kN, respectively. When compared to $F^r = 25$ kN, each actuating force value satisfies the performance specification. The uniform wear case performs slightly better than uniform pressure for small preference change to the right of the preference functions' peaks. However, both cases have an imprecision in the result on the order of the difference between F^r and the peak actuating force value. Thus the actuating force performance variable for either uniform wear or uniform pressure nominally satisfy the performance specification (at and to the left of the peak); yet, a small decrease in preference of one or more of the design variables may result in inadmissible performance values.

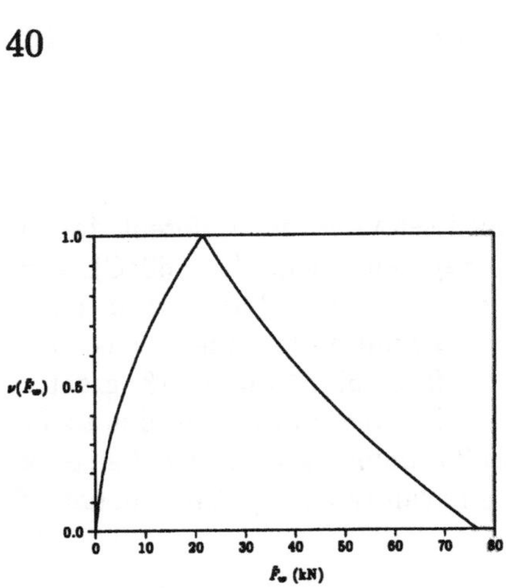

Actuating Force (Uniform Wear), $\tilde{F}_w$.

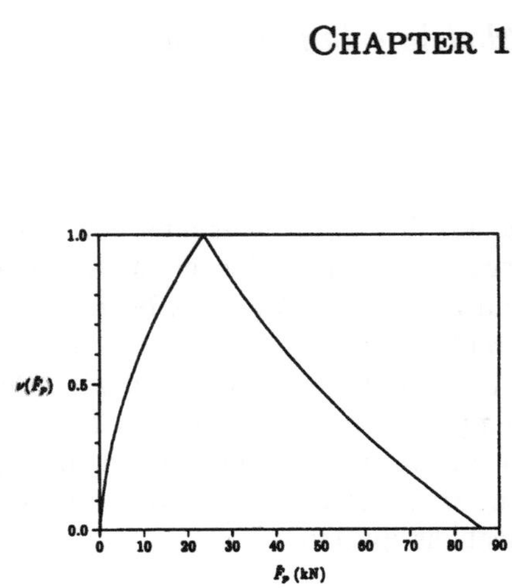

Actuating Force (Uniform Pressure), $\tilde{F}_p$.

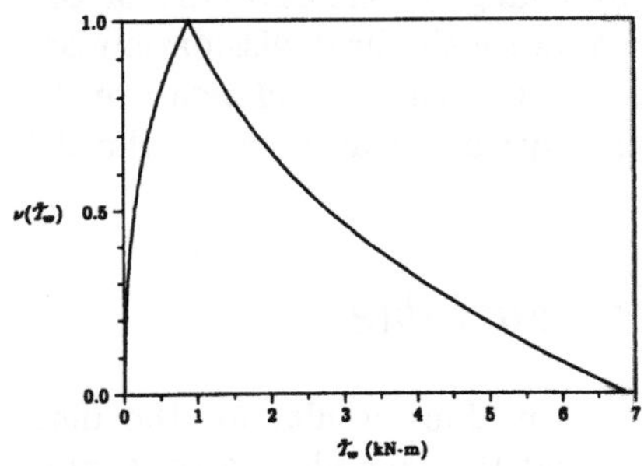

Torque (Uniform Wear), $\tilde{\mathcal{T}}_w$.

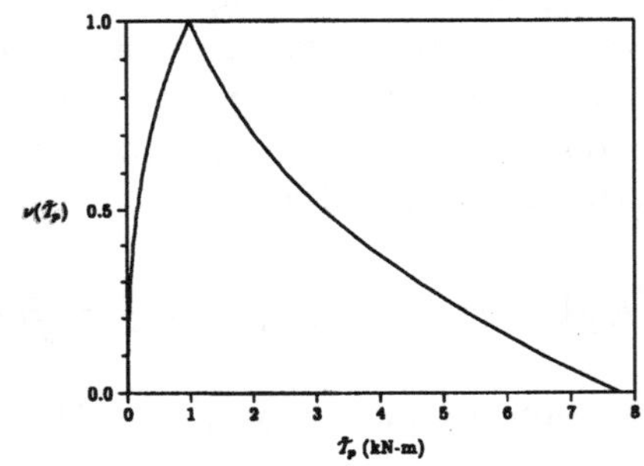

Torque (Uniform Pressure), $\tilde{\mathcal{T}}_p$.

Figure 15 Disk Brake: Induced Preferences (Force and Torque).

The induced preferences $\tilde{\mathcal{T}}_w$ and $\tilde{\mathcal{T}}_p$ are shown at the bottom of Figure 15. As in the drum brake configuration, Figure 14 represents the imprecise performance specification for torque. For preference equal to one (1), $\mathcal{T}^r = 0.579$ kN-m, $\mathcal{T}_w = 0.890$ kN-m, and $\mathcal{T}_p = 0.983$ kN-m. A comparison of $\mathcal{T}_w$ and $\mathcal{T}_p$ to $\mathcal{T}^r$ shows that both uniform wear and uniform pressure for the disk brake satisfy the torque design requirement. For small changes of the design variables (except M_v) to the left of the induced torques' peaks (e.g., a decrease in preference to 0.8), the values for $\mathcal{T}_w$ and $\mathcal{T}_p$ still satisfy the performance criteria. If the design variable M_v changes preference proportionally to any other input variable, the output torque values for uniform wear and uniform pressure will always satisfy the performance specification. Overall, the uniform pressure case performs better than uniform wear for torque considerations. Even though both cases have possible combinations of the design variables which do not meet the performance specification, satisfactory solutions do exist. Special care must be taken in adjusting M_v.

The top left graph in Figure 16 shows that the disk brake configuration satisfies the performance specification for temperature rise at the peak value ($\tilde{T}_{r\,(at\ \nu_{\tilde{T}_r}=1)}$ = 17.7°C). For decreasing preference to the right of the peak, large changes in preference (e.g., $\nu_{\tilde{T}_r}$ = 0.5) for any input variable still satisfy T_r^r. For repeated braking episodes, the results are similar to the drum brake system. Due to the large imprecision in $\tilde{T}_r$, a decrease in preference of any of the inputs to 0.8 or below may result in T_r values which may accumulate for successive braking actions, such that they approach the performance criteria, T_r^r. Therefore, T_r for the disk brake configuration is once again a critical performance variable, but not as critical as in the case of the drum brake since the values of $T_{r,disk}$ about the peak are smaller by a factor of two than $T_{r,drum}$ (e.g., $T_{r,disk}$ = 17.7°C compared with $T_{r,drum}$ = 35.3°C).

Figure 16 also contains the induced sets for the temperature decay profile $\tilde{T}_d$ for times of $t = 1$ sec and $t = 60$ sec. The peak value for the graph at the top right of Figure 16 corresponds to $\tilde{T}_{d\,(at\ \nu_{\tilde{F}_d}=1)}$ = 0.11°C, whereas the peak value for the graph at the bottom left of Figure 12 occurs at $\tilde{T}_{d\,(at\ \nu_{\tilde{F}_d}=1)}$ = 6.0°C. These results illustrate that the slow rate of temperature decay is generally similar to the drum brake configuration. Yet, for induced preference values about the peak of $\tilde{T}_d$, the magnitude of the temperature decay is significantly higher than the corresponding values for the drum brake.

To obtain a measure of the imprecision of the temperature decay in terms of the design variables (i.e., excluding T_r), T_r must be set to a crisp value as shown earlier for the drum brake. This requirement is because T_r is a performance variable, not a design variable, and the designer cannot select values for T_r. The results of the calculation, using constant T_r and imprecise design variables, may be found at the bottom right of Figure 16. Once again, T_r is the major contributor to the overall imprecision in $\tilde{T}_d$ (comparing the bottom two graphs in Figure 16), but the induced imprecision for the case of constant T_r is still significant, implying that a change in one or more of the design variables will result in a large change of temperature decay. In terms of the performance criteria T_d^r, the disk brake system dissipates heat to a greater extent than the results found for the drum brake system. Due to the possibility of large imprecision in the temperature decay profile for a corresponding change in the disk brake design variables, heat dissipation may be maximized beyond the values centered about the peak of the $\tilde{T}_d$ induced curve. The following

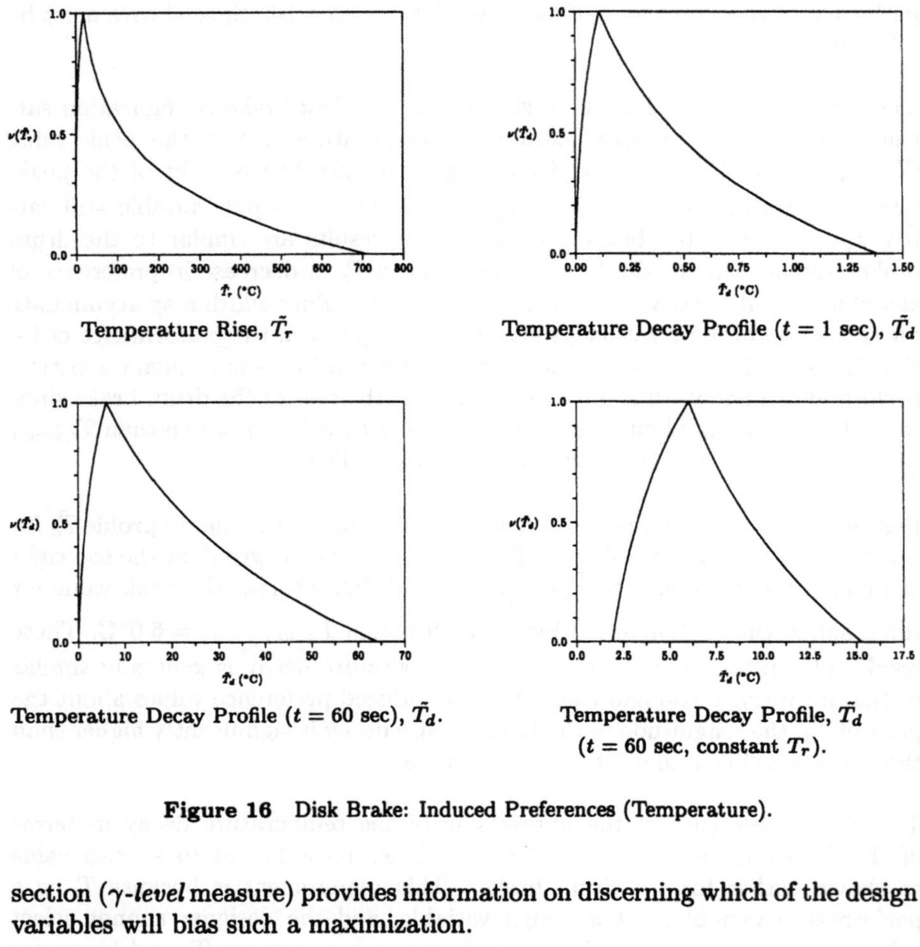

Temperature Rise, $\tilde{T}_r$

Temperature Decay Profile ($t = 1$ sec), $\tilde{T}_d$

Temperature Decay Profile ($t = 60$ sec), $\tilde{T}_d$.

Temperature Decay Profile, $\tilde{T}_d$ ($t = 60$ sec, constant T_r).

Figure 16 Disk Brake: Induced Preferences (Temperature).

section (γ-*level* measure) provides information on discerning which of the design variables will bias such a maximization.

6.5 Applying the γ-Level Measure (Sensitivity Analysis)

The γ-*level* measure may be used to provide the engineer with qualitative information concerning the role of the input variables in the brake design. The process of using and algorithmically implementing the γ-*level* measure may be found in [56].

The γ-*level* measure is a sensitivity measure, but in an unconventional mode. It measures the sensitivity of induced preference ($\nu(\vec{p})$) on the PVS to changes in design variable preference ($\mu(\vec{d})$). When a design variable is not related to a performance variable, the γ-*level* measure produces a zero (0.0) value. As the normalized γ-*level* measure increases in value, the corresponding design variable has a greater qualitative importance in determining the induced preference for the particular performance variable in question, meaning that a comparatively small change in the design variable preference produces a large change in performance preference. The γ-*level* measure can be used to order the design variables according to importance (sensitivity). Further, if one design variable has a significantly higher γ-*level* measure for an induced performance variable set, when compared to another design variable (where the second variable's measure is not zero), the designer may fix the value of the second variable without affecting performance (i.e., functional independence).

Drum Brake γ-Level Measure (Sensitivity Analysis)

Table 6 lists the results of the γ-*level* measure for the drum brake configuration as applied to each input variable with respect to each output set.

As shown in the third column of Table 6 for energy E, the γ-*level* measure indicates that the only significant design variables for energy considerations are the mass of the vehicle and the angular velocity ω. This result makes perfect sense when examining the terms of (1.35). When this equation is derived, all contributing factors to the energy are taken into account. The left portion of the equation contains the energy term due to the brake parts, while the right portion contains the contribution due to the kinetic energy of the vehicle. The energy of the brake parts will be insignificant when compared to the vehicle's kinetic energy. The γ-*level* measure indicates this result.

The γ-*level* measure results for the torque performance variable $\mathcal{T}$ are shown in the second column of Table 6. The design variables r, η, and p_a are shown to have nearly the same importance with respect to $\mathcal{T}$. Notice, however, that the design variable θ_1 contributes very little to the induced performance set. This indicates that θ_1 is essentially orthogonal to $\mathcal{T}$.

The conductivity design variable k has little contribution to T_d (a γ-*level* measure result of 0.004) and may be fixed to a representative material value. The γ-*level* measure is especially useful for determining the interactions between design variables and performance variables in this way.

	Performance Variables				
DVs	F_a	$\mathcal{T}$	E	T_r	T_d
a	0.21	0.0	0.0	0.0	0.0
r	1.0	0.92	0.0	0.44	0.0
θ_1	0.30	0.04	0.0	0.0	0.0
θ_2	0.43	0.41	0.0	0.0	0.0
η	0.77	0.92	0.0	0.0	0.0
p_a	0.94	1.0	0.0	0.0	0.0
c	0.93	0.0	0.0	0.0	0.0
b	0.64	0.0	0.0	0.0	0.0
ω	0.0	0.0	1.0	1.0	0.0
C	0.0	0.0	0.0	0.19	0.19
M_v	0.0	0.0	0.39	0.39	0.0
R_1	0.0	0.0	0.001	0.21	0.19
L_w	0.0	0.0	0.0	0.36	0.0
k	0.0	0.0	0.0	0.0	0.004
h	0.0	0.0	0.0	0.0	0.09
T_r	0.0	0.0	0.0	1.0	1.0

Table 6 γ-*level* **Measure Results for Drum Brake.**

Disk Brake γ-Level Measure (Sensitivity Analysis)

Table 7 lists the results of the γ-*level* measure for the disk brake configuration as applied to each input variable with respect to each output set.

The γ-*level* measure results for the torque performance variable ($\mathcal{T}_w$ for the uniform wear case, $\mathcal{T}_p$ for the uniform pressure case) are shown in the third and fourth columns of Table 7. The design variables r_0, r_i, η, and p_{max} or p_p are shown to have nearly the same importance with respect to $\mathcal{T}$. Notice, however, that the other design variables contribute very little to the induced performance set. This indicates that they are essentially orthogonal to $\mathcal{T}$.

Table 8 lists the γ-*level* results for the temperature decay performance variable where T_r takes on a representative crisp value. Since the temperature rise T_r can take on a wide range of values, depending on vehicle speed, etc., it is fixed to a nominal value here for comparison of the two alternative design configurations.

	Performance Variables					
DVs	F_w	F_p	$\mathcal{T}_w$	$\mathcal{T}_p$	T_r	T_d
p_{max}	1.0	0.0	0.95	0.0	0.0	0.0
r_o	0.55	0.49	1.0	0.99	0.37	0.0
r_i	0.93	1.0	0.82	0.96	0.0	0.0
η	0.0	0.0	0.82	0.96	0.0	0.0
p_p	0.0	0.98	0.0	1.0	0.0	0.0
ω	0.0	0.0	0.0	0.0	1.0	0.0
C	0.0	0.0	0.0	0.0	0.19	0.18
h	0.0	0.0	0.0	0.0	0.0	0.10
L_{w1}	0.0	0.0	0.0	0.0	0.28	0.27
M_v	0.0	0.0	0.0	0.0	0.39	0.0
T_r	0.0	0.0	0.0	0.0	1.0	1.0

Table 7 γ-*level* Measure Results for Disk Brake.

	Design Variables				
PV (constant T_r)	r_o	C	h	L_{w1}	T_r
T_d	0.0	0.69	0.35	1.0	0.0

Table 8 Disk Brake: γ-*level* Measure Results for T_d.

Because the induced results of $\tilde{T}_d$ for the disk brake (discussed above) indicate that a change in the design variables will have a large effect on the temperature decay, the most important design variables must be determined in order to focus the design efforts on those which will contribute the greatest to maximizing heat dissipation (i.e., satisfying T_d^r). Table 8 indicates that the design efforts should first focus on L_{w1} and subsequently C to produce better heat dissipation.

6.6 Discussion

The brake design example presented in this section demonstrates how imprecisely described configurations (design alternatives) may be evaluated with respect to a performance specification space. The alternative configurations may be compared for each dimension in the space in order to determine the major differences. It is seen (above) that both of the brake configurations satisfy the *actuating force* requirement where the induced performance variable preference $(\nu) = 1$. Although a small change in a design variable associated with the disk brake may make a large change in the actuating force performance variable so that it no longer satisfies its performance specification, the design variables associated with the actuating force performance variable are not ones the designer is likely to change. A discussion of the results above illustrates that the primary design variables which may require change from the peak values are L_{w1} and C (determined by the *γ-level measure* and the influence of those variables on the temperature rise). These design variables are independent of the actuating force performance variables for the disk brake, and will thus have no affect on the actuating force induced performance set.

The torque results for the disk brake configuration (also discussed above) are superior to those for the drum brake. Both cases have combinations of the design variables where the performance specification will not be satisfied. However, the disk brake is less sensitive to changes in the design variables when compared to the drum brake case.

The temperature rise performance variable is found to be a critical variable for both brake alternatives. Small changes in the design variables result in large changes in T_r, often beyond the acceptable range. In comparing the two alternatives, the disk brake configuration out-performs the drum brake by a factor of two in the region of interest. Care will need to be taken as the input variables are adjusted in the later design stages for either configuration due to the large degree of coupling between the temperature rise performance variable and the temperature decay profile performance variable.

The final dimension of the performance specification space to be considered is the temperature decay profile. The disk brake far out-performs the drum brake design. Not only do the results illustrate that heat is rejected faster (by a factor of six) for the disk brake, but they also show that better results may be obtained by adjusting the design variables relevant to the disk brake's T_d performance variable. This is not the case for the drum brake.

With these data, the engineer is equipped to choose between the disk and drum brakes. The only advantage of the drum brake for the given performance specifications and design variables is the small change in actuating force for comparatively large changes in design variables. As shown, the more important performance variables of temperature rise and temperature decay profile are handled much better by the disk brake design.

The brake design example illustrates the M$_o$I's set-based imprecision approach to preliminary design. Alternatives may be evaluated easily and efficiently by providing the designer with performance information and by reducing the usual complexities of working with the imprecise descriptions of the alternatives. Although this approach is constructed for use in the preliminary stages of design, it may also be useful later in the design process, especially when combined with other more traditional design tools.

7 CONCLUSION

Imprecise information is an important and intrinsic part of engineering design. However, this imprecision is often suppressed or overlooked, and rarely formalized. The *Method of Imprecision*, utilizing the mathematics of fuzzy sets, represents and manipulates imprecision for engineering design. This includes the effects of customer and designer preferences, formal strategies for trade-offs, iteration in engineering design, and noise (such as uncontrollable manufacturing and material property variations).

While other methods such as Utility Theory, Taguchi's method, etc., exist for similar purposes, they do not capture designers and customers preferences nor do they permit engineering trade-offs using a variety of strategies. A comparison of the M$_o$I with Utility Theory is more fully discussed in [35], and with other methods in [3].

Engineering design is nearly always iterative. When using the M$_o$I, preferences evolve through successive evaluations of imprecise design alternatives. Yet this is not traditional "point-by-point" design iteration: the methodology uses fuzzy sets representing preferences over many designs, providing more complete information earlier in the design process. This information can be propagated to downstream engineering groups, facilitating concurrent design. Set-based concurrent design [54] is a powerful paradigm that is enhanced by the use of fuzzy sets [3, 29].

Acknowledgements

This material is based upon work supported, in part, by: The National Science Foundation under a Presidential Young Investigator Award, Grant No. DMC-8552695; NSF Grant No. DDM-9201424, and a Sloan initialization grant from the Department of Mechanical Engineering at the Massachusetts Institute of Technology, funded in part by the DuPont Corporation. Additional support has been provided by a grant from the MIT Leaders for Manufacturing Research Program, a collaboration between MIT and many U.S. industrial firms. Any opinions, findings, conclusions or recommendations expressed in this publication are those of the authors.

REFERENCES

[1] AKAO, Y. *Quality Function Deployment: Integrating Customer Requirements into Product Design.* Productivity Press, Cambridge, 1990.

[2] ALGER, J., AND HAYS, C. *Creative Synthesis in Design.* Prentice Hall, Englewood Cliffs, 1964.

[3] ANTONSSON, E. K., AND OTTO, K. N. Imprecision in Engineering Design. *ASME Journal of Mechanical Design 117(B) (Special Combined Issue of the Transactions of the ASME commemorating the* 50th *anniversary of the Design Engineering Division of the ASME.)* (June 1995), 25–32. Invited paper.

[4] ASIMOW, M. *Introduction to Design.* Prentice-Hall, Inc., Englewood Cliffs, N. J., 1962.

[5] BELLMAN, R. E., AND GIERTZ, M. On the analytic formalism of the theory of fuzzy sets. *Information Sciences 5* (1973), 149–156.

[6] BIEGEL, P., AND PECHT, M. Design trade-offs made easy. *Concurrent Engineering 1*, 3 (May/June 1991), 29–40.

[7] COLTON, J., AND DESCANIO, J. An integrated, intelligent design environment. *Engineering with Computers 7* (1991), 11–22.

[8] COX, R. T. *The Algebra of Probable Inference.* Johns-Hopkins University Press, Baltimore, MD, 1961.

[9] DEGROOT, M. H. *Optimal Statistical Decisions.* McGraw-Hill, New York, 1970.

[10] DLESK, D. C., AND LIEBMAN, J. S. Multiple objective engineering optimization. *Engineering Optimization 6* (1983), 161–175.

[11] DONG, W. M., AND WONG, F. S. Fuzzy weighted averages and implementation of the extension principle. *Fuzzy Sets and Systems 21*, 2 (Feb. 1987), 183–199.

[12] DUBOIS, D., AND PRADE, H. *Fuzzy Sets and Systems: Theory and Applications.* Academic Press, New York, 1980.

[13] DUBOIS, D., AND PRADE, H. A class of fuzzy measures based on triangular norms. *International Journal of General Systems 8* (1982), 43–61.

[14] DUBOIS, D., AND PRADE, H. Criteria aggregation and ranking of alternatives in the framework of fuzzy sets. In *TIMS/Studies in the Management Sciences* (1984), Elsevier Science Publishers B.V., North-Holland, pp. 209–240.

[15] DUBOIS, D., AND PRADE, H. *Possibility Theory: An Approach to the Computerized Processing of Information.* Plenum Press, New York, 1988.

[16] ESCHENAUER, H., KOSKI, J., AND OSYCZKA, A., Eds. *Multicriteria Design Optimization.* Springer-Verlag, Berlin, 1990.

[17] FISHBURN, P. *Foundations of Expected Utility.* D. Reidel, 1982.

[18] FISHBURN, P. Normative theories of decision making under risk and under uncertainty. In *Non-Conventional Preference Relations in Decision Making*, J. Kacprzyk and M. Roubens, Eds. Springer-Verlag, New York, 1988, pp. 1–21.

[19] FISHBURN, P. C. Uncertainty aversion and separated effects in decision making under uncertainty. In *Combining Fuzzy Imprecision with Probabilistic Uncertainty in Decision Making*, J. Kacprzyk and M. Fedrizzi, Eds. Springer-Verlag, New York, 1988. Springer Lecture Notes in Economics and Mathematical Systems, vol. 310.

[20] FRENCH, S. *Decision Theory: An Introduction to the Mathematics of Rationality.* Ellis Horwood, Chichester, 1988.

[21] GAVIN, JR., J. G. Fly me to the Moon: An interview with Joseph G. Gavin, Jr. *Technology Review 97*, 5 (July 1994), 61–68.

[22] HALMOS, P. *Measure Theory.* Springer-Verlag, New York, 1974.

[23] HAUSER, J., AND CLAUSING, D. The house of quality. *Harvard Business Review* (May/June 1988), 63–73.

[24] HOELTZEL, D. A., AND CHIENG, W.-H. Systems for unified life-cycle mechanical engineering design: Shared-tool architectures versus distributed tool architectures. *Engineering with Computers 6*, 4 (1990), 211–222.

[25] HOLMES, M. F. Machine dynamics, The need for greater productivity. In *Research Needs in Mechanical Systems* (New York, 1984), K. N. Reid, Ed., ASME, pp. 140–159.

[26] KACPRZYK, J., AND ROUBENS, M., Eds. *Non-Conventional Preference Relations in Decision Making*, vol. 301 of *Lecture Notes in Economics and Mathematical Systems*. Springer-Verlag, New York, 1988.

[27] KEENEY, R., AND RAIFFA, H. *Decisions with Multiple Objectives: Preferences and Value Tradeoffs*. Cambridge University Press, Cambridge, U.K., 1993.

[28] KRANTZ, D., LUCE, R., SUPPES, P., AND TVERSKY, A. *Foundations of Measurement*, vol. I. Academic Press, New York, 1971.

[29] LAW, W. S., AND ANTONSSON, E. K. Hierarchical Imprecise Design with Weights. In *Proceedings of the Fourth IEEE International Conference on Fuzzy Systems (FUZZ-IEEE/IFES'95)* (Mar. 1995), vol. 1, IEEE, pp. 383–388.

[30] LAW, W. S., AND ANTONSSON, E. K. Optimization Methods for Calculating Design Imprecision. In *Advances in Design Automation - 1995* (Sept. 1995), vol. 1, ASME, pp. 471–476.

[31] OSYCSKA, A. *Multi-Criterion Optimization in Engineering with Fortran Examples*. Halstad Press, New York, 1984.

[32] OTTO, K. N. *A Formal Representational Theory for Engineering Design*. PhD thesis, California Institute of Technology, Pasadena, CA, June 1992.

[33] OTTO, K. N., AND ANTONSSON, E. K. Trade-Off Strategies in Engineering Design. *Research in Engineering Design 3*, 2 (1991), 87–104.

[34] OTTO, K. N., AND ANTONSSON, E. K. Trade-Off Strategies in the Solution of Imprecise Design Problems. In *Fuzzy Engineering toward Human Friendly Systems: Proceedings of the International Fuzzy Engineering Symposium '91, Volume 1* (Yokohama, Japan, Nov. 1991), T. Terano et al., Eds., LIFE, IFES, pp. 422–433.

[35] OTTO, K. N., AND ANTONSSON, E. K. The Method of Imprecision Compared to Utility Theory for Design Selection Problems. In *Design Theory and Methodology - DTM '93* (Sept. 1993), ASME, pp. 167–173.

[36] OTTO, K. N., AND ANTONSSON, E. K. Tuning Parameters in Engineering Design. *ASME Journal of Mechanical Design 115*, 1 (Mar. 1993), 14–19.

[37] Otto, K. N., and Antonsson, E. K. Design Parameter Selection in the Presence of Noise. *Research in Engineering Design 6*, 4 (1994), 234–246.

[38] Otto, K. N., Lewis, A. D., and Antonsson, E. K. Approximating α-cuts with the Vertex Method. *Fuzzy Sets and Systems 55*, 1 (Apr. 1993), 43–50.

[39] Otto, K. N., Lewis, A. D., and Antonsson, E. K. Determining optimal points of membership with dependent variables. *Fuzzy Sets and Systems 60*, 1 (Nov. 1993), 19–24.

[40] Papalambros, P., and Wilde, D. *Principles of Optimal Design*. Cambridge University Press, New York, 1988.

[41] Pugh, S. *Total Design*. Addison-Wesley, New York, 1990.

[42] Sandgren, E. A multi-objective design tree approach for the optimization of mechanisms. *Mechanisms and Machine Theory 25*, 3 (1990), 257–272.

[43] Scott, M. J., and Antonsson, E. K. Aggregation Functions for Engineering Design Trade-offs. In *9th International Conference on Design Theory and Methodology* (Sept. 1995), vol. 2, ASME, pp. 389–396.

[44] Shigley, J. E., and Mitchell, L. D. *Mechanical Engineering Design*, fourth ed. McGraw-Hill Book Company, New York, 1983.

[45] Siddall, J. N. *Analytical Decision Making in Engineering Design*. Prentice-Hall, 1972.

[46] Steuer, R. *Multiple Criteria Optimization: Theory, Computation, and Application*. J. Wiley, New York, 1986.

[47] Sugeno, M. *Theory of fuzzy integrals and its applications*. PhD thesis, Tokyo Institute of Technology, 1974.

[48] Sugeno, M. Fuzzy measures and fuzzy integrals – a survey. In *Fuzzy Automata and Decision Processes* (New York, 1977), M. Gupta et al., Eds., North-Holland, pp. 89–102.

[49] Sullivan, L. P. Quality function deployment. *Quality Progress* (June 1986), 39–50.

[50] Sveshnikov, A. *Problems in Probability Theory, Mathematical Statistics, and the Theory of Random Functions*. Dover Publications, New York, 1968.

[51] Tribus, M. *Rational Descriptions, Decisions, and Designs*. Pergamon Press, New York, 1969.

[52] ULLMAN, D. G. *The Mechanical Design Process.* McGraw Hill, New York, 1992.

[53] ULRICH, K. T., AND PEARSON, S. A. Does product design really determine 80% of manufacturing cost? Working Paper MSA 3601-93, MIT, Sloan School of Management, Cambridge, MA, Aug. 1993.

[54] WARD, A. C., LIKER, J. K., SOBEK, D. K., AND CRISTIANO, J. J. Set-based concurrent engineering and Toyota. In *Design Theory and Methodology - DTM '94* (Sept. 1994), vol. DE68, ASME, pp. 79–90.

[55] WHITNEY, D. E. Manufacturing by design. *Harvard Business Review 66*, 4 (July 1988), 83–91.

[56] WOOD, K. L., AND ANTONSSON, E. K. Computations with Imprecise Parameters in Engineering Design: Background and Theory. *ASME Journal of Mechanisms, Transmissions, and Automation in Design 111*, 4 (Dec. 1989), 616–625.

[57] WOOD, K. L., AND ANTONSSON, E. K. Modeling Imprecision and Uncertainty in Preliminary Engineering Design. *Mechanism and Machine Theory 25*, 3 (Feb. 1990), 305–324. Invited paper.

[58] WOOD, K. L., ANTONSSON, E. K., AND BECK, J. L. Representing Imprecision in Engineering Design – Comparing Fuzzy and Probability Calculus. *Research in Engineering Design 1*, 3/4 (1990), 187–203.

[59] WOOD, K. L., OTTO, K. N., AND ANTONSSON, E. K. A Formal Method for Representing Uncertainties in Engineering Design. In *First International Workshop on Formal Methods in Engineering Design* (Fort Collins, Colorado, Jan. 1990), P. Fitzhorn, Ed., Colorado State University, pp. 202–246.

[60] WOOD, K. L., OTTO, K. N., AND ANTONSSON, E. K. Engineering Design Calculations with Fuzzy Parameters. *Fuzzy Sets and Systems 52*, 1 (Nov. 1992), 1–20.

[61] WOODSON, T. *Introduction to Engineering Design.* McGraw-Hill, New York, 1966.

[62] YAGER, R. Fuzzy decision making including unequal objectives. *Fuzzy Sets and Systems 1* (1978), 87–95.

[63] ZADEH, L. A. Fuzzy sets. *Information and Control 8* (1965), 338–353.

2

MULTIPLE OBJECTIVE DESIGN OPTIMIZATION

Anoop K. Dhingra, Ph.D.

Assistant Professor of Mechanical Engineering
College of Engineering and Applied Science
University of Wisconsin, Milwaukee
Milwaukee, Wisconsin 53201, U.S.A.

First page of 35 pages of text of Chapter 2 (Page 53).

1 INTRODUCTION

Section 1, Page 53

2 PROBLEM STATEMENT

Section 2, Page 53

MULTIPLE OBJECTIVE DESIGN OPTIMIZATION

Anoop K. Dhingra
Department of Mechanical Engineering
University of Wisconsin, Milwaukee, WI 53201

1. INTRODUCTION

A common problem confronting a designer is selecting the best alternative from a set of feasible alternatives. In many situations this is not a difficult task as the best course of action is well defined, namely the alternative that minimizes (or maximizes) a well defined scalar-valued objective function. But in many decision making scenarios, no single objective function can adequately serve to compare the difference in desirability among feasible solutions. The theory of multiple objective optimization (MOO) deals with such problems and first arose in mathematical economics. In the framework of mathematical programming, the MOO problem was first mentioned by Kuhn and Tucker (1951) and in engineering literature by Zadeh (1963). For a comprehensive review of various aspects of the MOO problem, see Stadler (1984), Evans (1984), Stadler and Dauer (1992), and Koski (1993).

This chapter, addressing solution techniques for MOO problems, is organized as follows. Section 2 discusses the concept of a Pareto-optimal solution in the context of multiobjective problems. Several commonly used techniques for generating Pareto-optimal solutions to crisp MOO problems are discussed in section 3. Necessary theorems which insure Pareto-optimality of the solution are also outlined. Design problems where the objective functions and the constraints are not known precisely are addressed in section 4. The tools of fuzzy set theory are used to solve these ill-structured problems. The fundamental assumption in fuzzy optimization involving a use of linear membership functions is also examined in this section. Section 5 presents an algorithm for MOO based on the concept of cooperative fuzzy games. The approach presented is capable of solving problems with partly crisp and partly fuzzy objective functions, as well as partly crisp and partly fuzzy constraints. Numerical results obtained using fuzzy game theory for the design of high speed mechanisms, and the combined structural and control system optimization for flexible space structures are presented in section 6. Finally section 7 offers concluding remarks on various solution approaches to the MOO problem.

2. PROBLEM STATEMENT

A general MOO problem is of the form

$$\min \vec{f}(X) = [f_1(X), f_2(X), \ldots, f_k(X)]^T \tag{1}$$

subject to

$$X \in S = [X | X \in R^n, g_i(X) \le a_i, h_j(X) = b_j] \tag{2}$$

where X is an n-dimensional vector of variables, $f_1(X),..f_k(X)$ are k objective functions, and S is the set of feasible solutions. For a single objective problem, an optimum solution is defined as one that minimizes the objective function $f_i(X)$, $X \in S$. For a multiobjective problem, an attempt to define a vector minimal point as one at which all components of the objective function vector $\vec{f}$ are simultaneously minimized is not an adequate generalization since such 'utopia' points are seldom attainable. Therefore the concept of Pareto-optimality is frequently used in the context of MOO problems (Soland, 1979).

Definition 1: A solution $X^* \in S$ is Pareto-optimal if there is no $\bar{X} \in S$ such that $f_i(\bar{X}) \le f_i(X^*)$, $i=1,..,k$, and $f_{i_0}(\bar{X}) < f_{i_0}(X^*)$ for at least one $i_0 \in [1,..,k]$.

Definition 2: A feasible solution $X^* \in S$ is weakly Pareto-optimal if there exists no $\bar{X} \in S$ such that $f_i(\bar{X}) < f_i(X^*)$, $i=1,..,k$.

If X^* is Pareto-optimal, then X^* is weakly Pareto-optimal. The converse is true only under special circumstances. A further distinction of proper Pareto-optimality has been introduced by Geoffrion (1968) as

Definition 3: A feasible solution $X^* \in S$ is properly Pareto-optimal if X^* is Pareto-optimal and $\exists M > 0$ such that $\forall i$ and each $X \in S$ satisfying $f_i(X) < f_i(X^*)$, $\exists j$ such that $f_j(X) > f_j(X^*)$ and $[f_i(X^*) - f_i(X)]/[f_j(X) - f_j(X^*)] \le M$.

Solutions which are not properly Pareto-optimal are excluded from the decision set as they violate assumptions for a rational behavior. Further, unless an optimization problem is convex, only a locally optimum solution is guaranteed using standard nonlinear programming (NLP) techniques. Therefore, for a non-convex problem, the concept of Pareto-optimality is modified to introduce the notion of a locally Pareto-optimal solution as

Definition 4: A solution $X^* \in S$ is said to be locally Pareto-optimal if and only if $\exists \delta > 0$ such that X^* is Pareto-optimal in $S \cap N(X^*, \delta)$ where $N(X^*, \delta)$ denotes a neighborhood of X^*, i.e. the set $[X | X \in S, \|X - X^*\|_2 < \delta]$.

The solution set for Eqs. (1-2) is partially ordered and contains many Pareto-optimal solutions. One of these solutions is preferred by the decision maker (DM), and is determined through the use of a value function over the multiple objective functions of the problem. The value function transforms the MOO problem into one involving a single objective function. A maximization of the value function then yields the preferred (or compromise) solution.

3. CRISP APPROACHES FOR MULTIOBJECTIVE OPTIMIZATION

Most solution techniques for MOO involve a scalarization of the original multiobjective problem. The scalarization is done either by constructing a composite objective function, by considering the objective functions sequentially, or by defining an additional supercriterion function, $F: R^k \rightarrow R$. The theory of MOO uses these ideas, with various approaches differing in the manner in which Pareto-optimal solutions are generated, and the largely subjective nature of selecting a preferred (or compromise) solution from this set. The MOO techniques presented next require an apriori articulation of preference information from the DM, and serve as a starting point for developing the necessary and sufficient conditions for Pareto-optimality.

The objective functions of a MOO problem may be commensurable or non-commensurable. To have a common basis for comparison and avoid working with different objectives in different units, the objective functions $f_i(X)$ can be transformed into new objective functions F_i defined as $F_i=m_i f_i(X)$, $i=1,..,k$ (Dhingra et al. 1990b). Here the constant multipliers m_i are chosen so that for $i=1,..k$, $m_i f_i(X_0)=C$ at any feasible starting vector X_0. This scaling procedure insures that all objective functions are equal at a particular value of X_0. Hereafter, it will be assumed that the k objective functions (f_i) correspond to the k scaled objective functions (F_i) defined above. Further, it will be assumed that the nonlinear MOO problem given by Eqs. (1-2) is non-convex so that only a locally Pareto-optimal solution is guaranteed.

3.1 Global Criterion Formulation

This method requires no articulation of DM's preferences once the problem objectives and constraints have been defined. The DM is willing to accept whatever solution is obtained by minimizing some global criterion F(X) such as

$$\text{Find } X \in S \text{ which minimizes } F(X) = \sum_{i=1}^{k} \left[\frac{F_i(X) - F_i(X_i^*)}{F_i(X_i^*)} \right]^p \tag{3}$$

Here X_i^* is the optimum solution obtained by minimizing $F_i(X)$ with $X \in S$ and a value of p=2 is commonly used.

<u>Lemma 1</u>: For $1 \le p < \infty$, if X^* minimizes F(X), then X^* is Pareto-optimal.

Compromise solutions with $p = \infty$ correspond to the min-max criterion for which Pareto-optimality of the optimum solution is not guaranteed.

3.2 Utility Function Formulation

In this approach, the MOO is converted to

$$\text{Find } X \in S \text{ which maximizes } U(\vec{f}) \tag{4}$$

where $U(\vec{f})$ is the utility function of multiple objective functions. The utility function U can have many forms. The most common form assumes that U is additively separable with respect to all objective functions. Thus, if $U_i(F_i)$ is the utility function for the objective function F_i, the overall utility function U is defined as

$$U(\vec{F}) = \sum_{i=1}^{k} U_i(F_i) \tag{5}$$

An optimum solution X^* is found by maximizing $U(\vec{F})$ with $X \in S$. A special form of Eq. (5), known as the weighting method, is extensively used in MOO problems where $U = -\sum w_i F_i$. Here w_i denotes the relative importance of the objective function F_i. The MOO problem is now given as

$$\text{Find } X \in S \text{ which min } F(X) = \sum_{i=1}^{k} w_i F_i(X). \tag{6}$$

Lemma 2: For $w_i \geq 0$, if X^* minimizes F(X), then X^* is weakly Pareto-optimal. If the optimum solution is unique, then it is Pareto-optimal. Further for $w_i > 0$, if X^* minimizes F(X), then X^* is properly Pareto-optimal.
Lemma 3: If S is convex and each F_i is concave on S, then X^* is properly Pareto-optimal if and only if X^* solves Eq. (6) for some $w_i > 0$.

The main advantage of the utility function formulation is its simplicity. It is easier to assess k unidimensional utility functions (U_i's) than to assess $U(\vec{F})$ directly. Similarly, it is easier to get w_i's from the DM. The disadvantages of this approach are that there are only a few cases where the utility function is additively separable, and w_i depend not only on the achievement level of F_i but also the achievement level of F_i relative to F_j, $i \neq j$. Further, for non-convex problems, this approach may miss all but a finite number of Pareto-optimal solutions.

3.3 Goal Programming Method

In goal programming, there are two basic models: the Archimedian model and the preemptive model. The preemptive model deals with the generation of solutions whose criterion vectors are most closely related, in a lexicographic sense, to points in the utopian set. The Archimedian model, on the other hand, generates candidate solutions whose criterion vectors are closest, in a weighted L_p metric sense, to the utopian set in the criterion space. In Archimedian goal programming, a designer sets goals (b_i's) and relative weights (w_i's) for each of the objective functions (F_i's). An optimum solution $X^* \in S$ is defined as one that minimizes

$$F(X) = \left[\sum_{j=1}^{k} w_j \left(d_j^+ + d_j^- \right)^p \right]^{1/p}, \; p \geq 1 \text{ such that}$$

$$\begin{aligned} & F_j(X) + d_j^+ - d_j^- = b_j,\; j = 1,..,k \\ & d_j^-, d_j^+ \geq 0,\; j = 1,..,k \end{aligned} \tag{7}$$

where d_j^- and d_j^+ are the under- and over-achievement from the target goals for the jth objective function. If the goal vector is not chosen properly, there is no guarantee that the goal programming formulation will terminate at a Pareto-optimal solution. However, if the goals b_j are set equal to F_j^* obtained by minimizing F_j, an over-achievement of the goals b_j's is not possible. Consequently, d_j^+ need not be defined and the Pareto-optimality of the optimum solution is guaranteed.

Lemma 4: For $w_i > 0$, $b_i = F_i^*$ and $1 \le p < \infty$, if X^* is an optimal solution of the goal programming formulation, then X^* is Pareto-optimal.

3.4 Goal Attainment Method

The goal attainment formulation requires specification of goals b_i's and weights w_i's for the objective functions F_i's. An optimum solution X^* is determined by

$$\text{Find } X \in S \text{ which min } z \text{ such that } F_i(X) - w_i z \le b_i,\ i = 1,..,k. \tag{8}$$

Here z is a scalar variable unrestricted in sign and the weights w_i are normalized so that $\sum w_i = 1$. In the case of under-attainment of the desired goals, a smaller weighting coefficient is associated with more important objective functions. For an over-attainment of the desired goals, a smaller weighting coefficient is associated with less important objective functions.

The optimum solution obtained using the goal attainment formulation is fairly sensitive to the goal vector ($\vec{b}$) and the weighting vector (w) given by the DM. Depending upon the prescribed values of the goal vector, it is possible that the weighting vector w does not influence the optimum solution. Instead, the optimum solution X^* is determined by the nondominated solution nearest to $\vec{b}$.

Lemma 5: If X^* is Pareto-optimal, there exists a $(w, \vec{b})$ pair such that X^* is an optimum solution to Eq. (8). However, if X^* is an optimal solution to Eq. (8), then there is no guarantee that X^* is Pareto-optimal.

3.5 Bounded Objective Function Formulation

In this method, the minimum (L^i) and maximum (U^i) acceptable achievement levels for each objective function F_i are specified by the DM. An optimum solution is found by

$$\text{Find } X \in S \text{ which } \min F_r(X) \text{ such that } L^i \le F_i(X) \le U^i\ i = 1,..,k, i \ne r \tag{9}$$

This method, also referred to as the ε-constraint method, can be shown to lead to weakly Pareto-optimal solutions. However, if the optimum solution X^* to Eq. (9) is unique, then X^* is Pareto-optimal.

Lemma 6: $X^* \in S$ is Pareto-optimal if and only if $\forall r \in [1,..k]$, $\exists U^i \in R^{k-1}$ such that X^* is a unique optimum solution to Eq. (9).

For non-convex problems, by systematically varying L^i and U^i, the bounded objective formulation can generate the entire set of Pareto-optimal solutions. However, it may be noted that checking X^* for uniqueness is a difficult task. A further difficulty with this method is to prescribe values for L^i and U^i prior to any preliminary solution. Since the DM has to specify these values in an information void, this may result in a formulation with inconsistent constraints. Another question which needs to be addressed with this approach is which objective should be used as $F_r(X)$.

3.6 Lexicographic Method

In this method, the objectives are ranked in the order of importance by the DM. An optimum solution X^* is obtained by minimizing the objective functions, starting with the most important one and proceeding in the order of importance of the objectives. Let the subscripts of the objectives denote not only the objective function number, but also their priority. The solution procedure is given as follows

1. Starting with X_0, minimize $F_1(X)$ with $X \in S$. Let the optimum solution be X_1^* with $F_1^* = F_1(X_1^*)$. If X_1^* is unique, stop, otherwise go to the next step.
2. Starting from X_1^*, minimize $F_2(X)$ subject to $X \in S$ and an additional constraint of the form $0.95F_1^* \leq F_1(X) \leq 1.05F_1^*$. Let the optimum solution be X_2^*. If X_2^* is unique, stop, otherwise go to the next step.
3. Proceeding as outlined in the previous steps, the optimization problem at the *i*th step is given as: Starting from X_{i-1}^*, minimize $F_i(X)$ subject to the constraint set, $X \in S$ and *i-1* additional constraints of the form $0.95F_j^* \leq F_j(X) \leq 1.05F_j^*$ $j=1,..,i-1$. The procedure is terminated whenever a unique solution is obtained.

<u>Lemma 7</u>: If X^* is an optimum solution generated by the lexicographic method, then X^* is weakly Pareto-optimal. If X^* is unique, then it is Pareto-optimal.
For a problem involving *k* criteria, there are a total of *k!* ways in which the objective functions can be ranked by the DM. Since the solution obtained using the lexicographic method is fairly sensitive to objective function rankings, one should exercise caution using this method if some criteria are of nearly equal importance.

4. FUZZY APPROACHES FOR MULTIOBJECTIVE OPTIMIZATION

The classical crisp optimization techniques discussed in the previous section can be used only when the MOO problem is well defined. This requires that all the problem data be known precisely, the constraints delimit a well defined set of feasible solutions, and the objective functions be well defined and easy to formulate. Frequently while modeling real world design problems, the problem data, constraints and/or objective functions cannot be stated precisely. Consequently a DM is unable to state the problem in precise mathematical terms. This, in turn, limits the applicability of crisp MOO techniques. Fuzzy set theories help alleviate such modeling difficulties by enabling a DM to structure and describe activities which differ from each other vaguely, to formulate them in models, and to use these models for problem solving and decision making (Zadeh, 1965 and 1973).

The first application of fuzzy theories to decision making processes was presented by Bellman and Zadeh (1970). An application of fuzzy optimization techniques to linear programming problems with single and multiple objectives functions has been presented by Zimmermann (1978). The concept of efficient and weakly efficient solutions in the context of fuzzy MOO problems is discussed by Feng (1983). Wang and Wang (1985) have used the method of level cut solutions for the fuzzy optimum design of structures. An application of fuzzy optimization techniques to the design of mechanical and structural systems has been discussed by Rao (1987a, b). Additional applications of fuzzy theories to multiobjective design problems are given in Dhingra et al. (1990a, 1992a, 1992b), Dhingra and Rao (1991), Dhingra and Moskowitz (1991), and Kim et al. (1991). For a comprehensive survey of fuzzy optimization, see Luhandjula (1989) and Zimmermann (1986 and 1992).

4.1 Symmetrical Models for Fuzzy Optimization

Traditional schemes for design optimization assume that all the problem data are known precisely, that the constraints delimit a well defined set of feasible decisions, and that the objective functions are well defined and are easy to formulate. An optimum decision is that combination of decision variables X^* which results in the 'highest degree of satisfaction' for the objective functions $\vec{f}$. For a problem involving uncertainty and fuzziness in the input design data, this notion of optimization needs to be modified. In this ill-structured decision making scenario, both the objective functions and the constraints can be represented by fuzzy sets which, in turn, are characterized by their membership functions.

4.1.1 Lambda Formulation

Consider again the MOO problem given by Eqs. (1-2). If the DM does not exhibit a *minimizing* behavior but acts as a *satisfier*, the MOO problem can be stated as

$$\text{Find } X \in S \text{ such that } f_i(X) \le a_i,\ i = 1,..k. \tag{10}$$

Here a_i denotes the aspiration levels for the k objective functions. The satisfaction level of the DM increases or decreases monotonically as the design approaches the aspiration levels depending on whether they represent the upper or lower bounds of acceptability. If f_i^{min} and f_i^{max}, respectively represent the lower and upper bounds of aspirations with respect to f_i, then the DM does not accept a solution for which $f_i \ge f_i^{max}$ and is completely satisfied whenever $f_i \le f_i^{min}$. Thus one can express the objective functions f_i by fuzzy sets whose membership functions decrease monotonically from 1 at f_i^{min} to 0 at f_i^{max}. Eq. (10) can be rewritten as

Find $X \in S$ such that $f_i(X) \tilde{\in} \overline{F}_i(X)$, $i = 1,..,k$. (11)

Here $\overline{F}_i$ denotes the allowable interval for the fuzzy objective function f_i, $\overline{F}_i = [f_i^{\min}, f_i^{\max}]$. The overbar indicates the presence of fuzzy information. If a min-operator is employed for aggregating fuzzy sets, an optimum solution to the fuzzy optimization problem (Eq. 11) is obtained by

Find $X \in S$ and λ which max λ such that $\lambda \le \mu_{\overline{F}_i}(X)$, $i = 1,..,k$ (12)

Here $\mu_{\overline{F}_i}$ are the membership functions of fuzzy objective functions f_i, denoted by fuzzy sets $\overline{F}_i$. Eq. (12) is known as the lambda approach to fuzzy optimization.

Theorem 1: If X^* is an optimum solution to Eq. (12), then any $\overline{X} \in S$ that strictly dominates X^* is also an optimum solution to Eq. (12).

Theorem 2: If X^* is a unique optimum solution to Eq. (12), then X^* is a Pareto-optimal solution to the MOO problem given by Eqs. (1-2).

Proof: The proofs for both these theorems are given in Dhingra et al. (1992a).

Consider next a generalization of Eq. (10) when not only the objective functions are fuzzy, but the constraints also are fuzzy. This problem can be stated as

Find X such that

$$f_i(X) \tilde{\in} \overline{F}_i,\ i = 1,..,k,\ g_j(X) \tilde{\in} \overline{G}_j,\ j = 1,..,nfg,\ h_j(X) \tilde{\in} \overline{H}_j,\ j = 1,..,nfh. \quad (13)$$

Here $\overline{F}_i, \overline{G}_j$ and $\overline{H}_j$ denote the allowable tolerance intervals for the fuzzy goals (f_i) and the fuzzy constraints (g_j and h_j). The fuzzy constraint $g_j \tilde{\in} \overline{G}_j$ indicates that g_j is a member of $\overline{G}_j$ such that $\mu_{\overline{G}_j}(g_j(X)) > 0$, where $\mu_{\overline{G}_j}$ is the membership function of the fuzzy set $\overline{G}_j$. A fuzzy feasible region is defined by considering all fuzzy constraints as

$$\overline{R}(X) = \left[\bigcap_{j=1}^{nfg} \overline{G}_j(X)\right] \cap \left[\bigcap_{j=1}^{nfh} \overline{H}_j(X)\right] \quad (14)$$

This gives the overall degree of satisfaction of alternative X with respect to all fuzzy constraints. An alternative X is considered feasible provided $\mu_{\overline{R}}(X) > 0$. A fuzzy decision is defined as the confluence of the fuzzy goals ($\overline{F}_i$) and the fuzzy constraints ($\overline{G}_j$ and $\overline{H}_j$) as

$$\overline{D}(X) = \overline{F}_1 \cap \ldots \cap \overline{F}_k \cap \overline{G}_1 \cap \ldots \cap \overline{G}_{nfg} \cap \overline{H}_1 \cap \ldots \cap \overline{H}_{nfh} \quad (15)$$

In terms of the membership values of the fuzzy objective functions and the fuzzy constraints, and assuming that the logical *and* (min) corresponds to an intersection, the resulting decision is

$$\mu_{\overline{D}}(X)=\left[\bigcap_{i=1}^{k}\mu_{\overline{F}_i}(X)\right]\cap\left[\bigcap_{j=1}^{nfg}\mu_{\overline{G}_j}(g_j(X))\right]\cap\left[\bigcap_{k=1}^{nfh}\mu_{\overline{H}_k}(h_k(X))\right] \text{ or} \tag{16}$$

$$\mu_{\overline{D}}(X)=\min_{i,j,k}\left[\mu_{\overline{F}_i}(X),\mu_{\overline{G}_j}(X),\mu_{\overline{H}_k}(X)\right] \tag{17}$$

Because of the symmetry of this aggregation procedure with respect to fuzzy goals and fuzzy constraints, there is no longer any distinction between the goals and the constraints of a decision process. An optimum solution X^* is one at which the membership function of $\overline{D}$ attains its maximum, i.e.

$$\text{Find } X \in \overline{D} \text{ which max } \mu_{\overline{D}}(X) \tag{18}$$

A sufficient condition for a unique maximum is that $\overline{D}$ be strongly convex set, i.e., $\overline{D}$ is convex with a unimodal membership function.

Computational Procedure for Symmetrical Models

A optimum solution to fuzzy problems given by Eqs. (12) and (18) is determined by (i) finding solutions to the single objective optimization problems, (ii) determining the best and worst values for each objective function, (iii) using these values as the boundaries of the fuzzy ranges for the fuzzy objective functions in the optimization problem, and (iv) solving the resulting fuzzy optimization problem.

A linear membership function of a fuzzy objective function is constructed as

$$\mu_{\overline{F}_i}(X)=\begin{cases}0, & \text{if } f_i(X)\ge f_i^{\max}\\ \left(\dfrac{f_i^{\max}-f_i(X)}{f_i^{\max}-f_i^{\min}}\right) & \text{if } f_i^{\min}<f_i(X)<f_i^{\max},\ i=1,..,k\\ 1, & \text{if } f_i(X)\le f_i^{\min}\end{cases} \tag{19}$$

where $f_i^{\min}=\min_j f_i(X_j^*)$ and $f_i^{\max}=\max_j f_i(X_j^*)$, and X_j^* is the optimum solution for the jth objective function. For fuzzy inequality constraints stated as $g_j \le a_j + c_j$, $j=1,..nfg$ where c_j denotes the distance by which the boundary of the jth constraint is moved, a linear membership function for the jth inequality constraint is given as

$$\mu_{\overline{G}_j}(X)=\begin{cases}0, & \text{if } g_j(X)\ge a_j+c_j\\ 1-\left(\dfrac{g_j(X)-a_j}{c_j}\right), & \text{if } a_j<g_j(X)<a_j+c_j,\ j=1,..,nfg\\ 1, & \text{if } g_j(X)\le a_j\end{cases} \tag{20}$$

For a fuzzy equality constraint stated as $h_j(X)=b_j \pm d_j$, $j=1,..,nfh$ where d_j denotes the distance by which the boundary of the jth constraint is moved, a linear membership function for the jth equality constraint is constructed as

$$\mu_{\overline{H}_j}(X)=\begin{cases}0, & if\ h_j(X)\geq b_j+d_j\\ 1-\left(\dfrac{|h_j(X)-b_j|}{d_j}\right) & if\ b_j-d_j<h_j(X)<b_j+d_j,\ j=1,..,nfh\\ 0, & if\ h_j(X)\leq b_j-d_j\end{cases}\quad(21)$$

Once the membership functions of the fuzzy objective functions and the fuzzy constraints, i.e. $\mu_{\overline{F}_i}$, $\mu_{\overline{G}_j}$ and $\mu_{\overline{H}_j}$ are known, Eq. (18) can be posed as

Find X and λ which max λ such that

$$\begin{aligned}&\lambda\leq\mu_{\overline{F}_i},\ i=1,..k\\ &\lambda\leq\mu_{\overline{G}_i},\ i=1,..,nfg\\ &\lambda\leq\mu_{\overline{H}_i},\ i=1,..,nfh\\ &X\in S=[X|X\in R^n,g_i\leq a_i,h_j=b_j,i=1,..,ncg,j=1,..nch]\end{aligned}\quad(22)$$

Eq. (22), which is a generalization of Eq. (18), has *ncg*-crisp and *nfg*-fuzzy inequality constraints, and *nch*-crisp and *nfh*-fuzzy equality constraints. Further, Eq. (22) is more general compared to Eq. (12) because of the presence of fuzzy constraints. Whereas Eq. (12) has only fuzzy objectives and crisp constraints, Eq. (22) has fuzzy objectives and constraints which are partly crisp and partly fuzzy. Thus, the concept of Pareto-optimality used for crisp optimization problems needs to be revised to introduce the concept of a fuzzy Pareto-optimal solution. This is done because the solutions differ not only with respect to objective function values, but also with respect to their degree of feasibility (Zimmermann, 1991).

Definition 5: Let $f_i: R^n\rightarrow R$, $i=1,..,k$ be the objective functions and $\mu_{\overline{G}_i}: R^n\rightarrow[0,1]$, $i=1,..,nfg$ and $\mu_{\overline{H}_j}: R^n\rightarrow[0,1]$, $j=1,..,nfh$ be the membership functions of fuzzy constraints. A solution $X^*\in S$ is said to be fuzzy Pareto-optimal if and only if there exits no $\overline{X}\in S$ such that

$$\begin{aligned}&f_i(\overline{X})\leq f_i(X^*)\ \forall i\in[1,..,k]\\ &\text{and } \mu_{\overline{G}_i}(\overline{X})\geq\mu_{\overline{G}_i}(X^*)\ \forall i\in[1,..,nfg]\\ &\text{and } \mu_{\overline{H}_i}(\overline{X})\geq\mu_{\overline{H}_i}(X^*)\ \forall i\in[1,..,nfh]\\ &\text{and } \exists j\in[1,..,k]\ni f_j(\overline{X})<f_j(X^*)\\ &\text{or } \exists j\in[1,..,nfg]\ni\mu_{\overline{G}_j}(\overline{X})>\mu_{\overline{G}_j}(X^*)\\ &\text{or } \exists j\in[1,..,nfh]\ni\mu_{\overline{H}_j}(\overline{X})>\mu_{\overline{H}_j}(X^*)\end{aligned}\quad(23)$$

It can be shown that if there exists a unique optimal solution to Eq. (22), then this solution is fuzzy Pareto-optimal.

4.2 Nonsymmetrical Models for Fuzzy Optimization

The lambda-formulation and its variants presented above are based on the assumption that the objective functions and/or the constraints can be modeled as fuzzy sets, and the degree of membership of solutions to objective functions and constraints can be considered compatible. In certain design processes, the objective functions cannot be represented by fuzzy sets. Such situations arise when the DM exhibits an optimizing (rather than satisficing) behavior with respect to the objective functions, but is unable to describe the solution space precisely. In this case, the goals and the constraints do not play a symmetric role, and one is faced with the problem of determination of an extremum of crisp objective functions over a fuzzy domain. Two approaches which can be used to solve such problems are discussed next.

4.2.1 Alpha-cut Approach

In the alpha-cut approach, the fuzzy optimization problem is transformed into a crisp optimization problem as follows

Find $X \in S$ which min f(X) such that

$$\begin{aligned} &\alpha \le \mu_{\overline{G}_i},\ i = 1,..nfg \\ &\alpha \le \mu_{\overline{H}_j},\ j = 1,..nfh \end{aligned} \tag{24}$$

By computing all alpha-cuts of the solution space and the corresponding objective function values, the fuzzy decision set consists of the optimum values of the objective function with a membership value equal to alpha. Eq. (24) can also be generalized to include multiple objective functions as

Find $X \in S$ which min $F(\vec{f}(X))$ such that

$$\begin{aligned} &\alpha \le \mu_{\overline{G}_i},\ i = 1,..nfg \\ &\alpha \le \mu_{\overline{H}_j},\ j = 1,..nfh \end{aligned} \tag{25}$$

Here $F: R^k \to R$ can, for instance, be defined as $F(X) = -\prod_{i=1}^{k}[F_{iu} - f_i]$ with $F_{iu} = f_i^{max}$. This functional form of F corresponds to the bargaining model used by the players when the MOO problem is treated as a cooperative game. The F_{iu} values can be determined using the computational procedure outlined earlier. The scalarization techniques for MOO presented in section 3 can also be used to construct F.

4.2.2 Maximizing Decision Approach

For the alpha-cut approach, the optimum solution is a fuzzy decision set with optimum solutions differing with respect to the degree of membership of the solution space. Frequently in engineering design problems, a unique solution is desired for only one of a multitude of solutions obtained using the alpha-cut approach can be implemented. The maximizing decision approach, based on the concept of a maximizing set, helps determine the preferred solution.

Let $f: R^n \rightarrow R$ be bounded from below by f^* and from above by f^0. The maximizing set $\overline{M}_f$ is defined by its membership function $\mu_{\overline{M}_f} = (f^0 - f(X)) / (f^0 - f^*)$.
Further, let $\overline{C}$ be the fuzzy feasible region, $S(\overline{C})$ be the support of $\overline{C}$ and C_1 be the alpha-cut of $\overline{C}$ for $\alpha = 1$. Then the membership function of the objective function f is defined as

$$\mu_{\overline{F}}(X) = \begin{cases} 1, & \text{if } f(X) \le \min_{S(\overline{C})} f \\ \left(\dfrac{\min_{C_1} f - f(X)}{\min_{C_1} f - \min_{S(\overline{C})} f} \right) & \text{if } \min_{S(\overline{C})} f < f(X) < \min_{C_1} f \\ 0, & \text{if } f(X) \ge \min_{C_1} f \end{cases} \tag{26}$$

An optimum solution to the fuzzy optimization problem is now obtained as

Find $X \in S$ and λ which max λ such that

$$\begin{aligned} &\lambda \le \mu_{\overline{F}} \\ &\lambda \le \mu_{\overline{G}_i}, i = 1,..,nfg \\ &\lambda \le \mu_{\overline{H}_i}, i = 1,..,nfh \end{aligned} \tag{27}$$

4.3 Nonlinear Membership Functions

A major assumption underlying solution to fuzzy optimization problems involves a use of linear membership functions for all fuzzy sets. The linear approximation is commonly used because of its simplicity and is defined by fixing two points, the upper and lower levels of acceptability. If fuzzy set theory is to be considered a purely formal theory, such an assumption is acceptable, even though some kind of formal justification of this assumption would be desirable. If, however, fuzzy set theory is used to model real decision making processes, and an assertion is made that the resulting models are true models of reality, then some kind of empirical justification for this assumption is necessary. In view of this, several nonlinear shapes for membership functions, such as concave or convex shaped membership functions were analyzed by Dhingra and Moskowitz (1991) and Dhingra et al. (1992a) to determine their impact on the design process. The marginal rate of increase (or decrease) of membership values as a function of design parameter values is not constant for these nonlinear membership functions, as is the case with linear membership functions. These nonlinear shapes offer potential benefits in terms of realism and were chosen consistent with varying perceptions of the decision maker.

The nonlinear membership functions were classified by Dhingra et al. (1992a) based on the coefficient of membership satiation, m(X) defined as

$$m(X) = \mu''(X) \tag{28}$$

where $\mu''(X)$ is the second derivative of the membership function. This definition of m(X) is analogous to the Arrow-Pratt measure of risk aversion and the Dyer-Sarin (1982) measure of value satiation used in decision analysis for characterizing utility and measurable value functions respectively. It may be noted that this definition of m(X) does not include $\mu'(X)$ because a linear transformation of membership functions is not possible, which is the case with utility or value functions. A positive value of m(X) corresponds to increasing marginal membership values at a given value of X (convex functions). Similarly a negative value of m(X) corresponds to a decreasing marginal membership values (concave functions), and m(X)=0 is equivalent to constant marginal membership values (linear functions). Second order effects which determine whether m(X) is increasing, constant or decreasing over the parameter range of interest, while retaining its sign, are also considered in Dhingra et al. (1992a). The sine and exponential ($k > 0$) functions were used to model increasing and decreasing values of m(X) over the range of definition $m(X) > 0$. The logarithmic, quadratic, and exponential ($k < 0$) functions were used to model increasing, constant and decreasing values of m(X) when the membership satiation coefficient is negative. It was seen that the final design is strongly influenced by the sign of m(X). Second order effects which govern whether m(X) is increasing, decreasing, or constant over the range of interest were found not influence the results significantly.

4.4 MODES: A Tool for Multiple Objective Design Optimization

The MOO techniques presented in sections 3 and 4, as well as several other state-of-the-art algorithms, have been implemented into a general purpose optimization program for solving multiple objective engineering design problems. MODES (Multiple Objective Design of Engineering Systems) is a FORTRAN program versatile enough to solve both crisp and fuzzy optimization problems with single as well as multiple objective functions (Kumar et al. 1990). It provides a general, computer-aided design methodology for solving a wide array of engineering design problems. Eighteen MOO techniques have been implemented in the program to accommodate differing design applications. An array of options are also available for solving the single objective optimization problem. A separation of analysis and optimization modules facilitates an integration of other analysis programs with MODES with relative ease. The modular structure of MODES also maintains enough flexibility to enhance the code as the state-of-the-art advances.

5. COOPERATIVE FUZZY GAMES

The applicability of cooperative game theory in a design process is examined next. The concepts in cooperative game theory and fuzzy set theory are combined to yield an optimization approach referred to, herein as, cooperative fuzzy games. The concept of cooperative fuzzy games can be applied to solve not only well- and ill-structured single and MOO problems, but also design problems with no objective functions. A formulation capable of solving problems with partly crisp and partly fuzzy objective functions, as well as partly crisp and partly fuzzy constraints is presented. It is shown that existing techniques for solving crisp and fuzzy optimization problems are special cases of the general formulation.

5.1 The Bargaining Model

Game theory is a mathematical tool for the analysis of conflict of interest situations. The conflict is comprised of players who select strategies from a set of available strategies. While the conflicting payoffs may put the players at cross purposes, there may also be room for cooperation between them. Game theory abstracts essential elements of such competitive situations, put them in mathematical models, and uses these models for decision making (Nash, 1950 and Leitmann, 1975).

In the game theoretic approach, a MOO problem is viewed as a cooperative game consisting of *k* players where each player corresponds to an objective function to be minimized. The players engage in bargaining to select a mutually beneficial strategy such that their distance from the worst objective function value is maximized. The bargaining function B(X) which helps determine a compromise solution amongst the *k*-objective functions is defined as

$$B(X) = \prod_{i=1}^{k} [F_{iu} - f_i(X)]^{w_i} \tag{29}$$

Here w_i is the relative degree of importance of objective function f_i and F_{iu} denotes the worst value of f_i that the designer is willing to accept. The worst values F_{iu} are determined by solving *k* single objective optimization problems and by finding

$$F_{iu} = \max_{\substack{j=1,\ldots,k \\ i \neq j}} f_i(X_j^*) \tag{30}$$

where X_j^* denotes the optimum solution of the *j*th objective function when solved as a single objective optimization problem. Thus the game theoretic formulation for a multiobjective problem yields

$$\text{Find } X \in S^* \subset S \text{ which max B(X).} \tag{31}$$

Here $S^* = [X | X \in S, (F_{iu} - f_i(X)) > 0]$.

Theorem 3: If X^* is an optimum solution to Eq. (31), then X^* is Pareto-optimal.

Proof: See Dhingra et al. (1992b).

Game theoretic class of methods for MOO (a) yield a solution which is Pareto-optimal, (b) require minimal amount of subjective information from the DM, (c) require no interpersonal comparison of utilities, and (d) insure that at the final solution all objective functions attain acceptable values. Most MOO techniques presented in section 3 lack one or more of these desirable features which cooperative game theory possesses. However, this formulation can be used only when the MOO problem is well defined. A generalization of Eq. (31) when the problem data, constraints or objective functions are not known precisely is presented next. To model the imprecision in the problem formulation, the tools of fuzzy set theory are used.

5.2 General Design Problem

Consider next a general version of the 'crisp' MOO problem given by Eqs. (1-2). The general version has *ncf* crisp and *nff* fuzzy objective functions, *ncg* crisp and *nfg* fuzzy inequality constraints, and *nch* crisp and *nfh* fuzzy equality constraints. A Pareto-optimal solution to this MOO problem is obtained by solving the following problem which combines fuzzy and game theoretic techniques presented earlier (Dhingra and Rao 1995).

$$\max\ B + a\lambda$$

subject to

$$\begin{aligned} &\lambda \le \mu_{\tilde{F}} \\ &\lambda \le \mu_{\tilde{G}_i}\ , i = 1,..,nfg \\ &\lambda \le \mu_{\tilde{H}_i}\ , i = 1,..,nfh \\ &X \in S = [X | X \in R^n, g_i \le a_i, h_j = b_j, i = 1,..,ncg, j = 1,..nch] \end{aligned} \tag{32}$$

where B and $\mu_{\tilde{F}}$ are defined as

$$B = \prod_{i=1}^{ncf} [F_{iu} - f_i], \quad \mu_{\tilde{F}} = \left[\prod_{i=1}^{nff} \mu_{\tilde{F}_i} \right]^{1/nff} \tag{33}$$

and a is a DM supplied parameter. The crisp objective functions are combined using the bargaining function B given by Eq. (33). In this equation, F_{iu} denotes the worst (status-quo) value for the *i*th objective function, determined using Eq. (30). This value helps each player analyze the maximum possible benefit he can obtain while negotiating with other players. It also apprises each player i of the extreme conditions of other players $[f_j(X_j^*), F_{ju}]$, helps him in guaranteeing against his worst value (F_{iu}), and insures that none of the players begin bargaining from a reference value which is unacceptable to other players.

The membership functions of fuzzy equality ($\mu_{\tilde{H}_i}$) and inequality ($\mu_{\tilde{G}_i}$) constraints are aggregated using the min operator. The min operator does not permit any tradeoff between the fuzzy constraints which must all be satisfied during the optimization process. To preserve the cooperative attitude expressing a tradeoff between fuzzy goals (Dhingra and Rao, 1995), the geometric mean operator is used for aggregating membership functions corresponding to fuzzy objective functions (Eq. 33). Since the complete design problem entails that both the fuzzy goals *and* the fuzzy constraints be satisfied, a min operator is used to aggregate fuzzy sets corresponding to fuzzy objective and constraint functions. The crisp equality and inequality constraints are included in the problem formulation as given in Eq. (32).

Eq. (32) can be used to solve not only well- and ill-structured single and MOO problems, but also preliminary decision making and design problems with no objective functions. Depending on the values for *nff, ncf, ncg, nfg, nch* and *nfh*, it can be shown that existing techniques for solving crisp and fuzzy optimization problems are special cases of the fuzzy game theoretic formulation (Eq. 32). Ten special cases along with necessary theorems for optimality are outlined next.

Case 1: nff=nfg=nfh=0, ncf=1. The proposed formulation reduces to

$$\text{Find } X \in S \text{ which } \max\,(F_{1u} - f_1) \qquad (34)$$

or in other words: find $X \in S$ which minimizes f_1. This is a mathematical programming problem with a single objective function and crisp constraints. An optimum solution to this problem is obtained using standard optimization techniques.

Case 2: nff=nfg=nfh=0, $ncf \geq 2$. The proposed model reduces to

$$\text{Find } X \in S \text{ which max B(X).} \qquad (35)$$

This formulation is identical to the game theoretic formulation given by Eq. (31).

Case 3: nff=ncg=nch=0, $ncf \geq 2$. The proposed formulation yields

$$\max B + a\lambda$$

subject to

$$\lambda \leq \mu_{\bar{G}_i},\, i = 1,..nfg$$

$$\lambda \leq \mu_{\bar{H}_j},\, j = 1,..nfh \qquad (36)$$

The functional form of the objective function incorporates a tradeoff between an improvement in the value of the bargaining function (B) and the degree of constraint violation (1-λ). When λ=1, all fuzzy constraints are completely satisfied. The value of the bargaining function is lowest for this case. When all fuzzy constraints are completely violated (λ=0) and B attains its highest value. Therefore the composite function B + a λ achieves a tradeoff between an improvement in the value of the bargaining function (B) and the degree of constraint violation.

If $ncg \geq 0$ and $nch \geq 0$, the previous formulation can be generalized to yield

$$\text{Find } X \in S \text{ which } \max B + a\lambda$$

subject to

$$\lambda \leq \mu_{\bar{G}_i},\, i = 1,..nfg$$

$$\lambda \leq \mu_{\bar{H}_j},\, j = 1,..nfh \qquad (37)$$

Optimization problems with crisp objective functions and crisp/fuzzy constraints can alternately be cast in the following form:

Find $X \in S$ which max B such that

$$\alpha \le \mu_{\overline{G}_i},\ i = 1,..nfg$$
$$\alpha \le \mu_{\overline{H}_j},\ j = 1,..nfh \tag{38}$$

This formulation is a generalization of the alpha-cut approach for fuzzy optimization that allows for multiple objective functions. It can be used to find an extremum of crisp objective function(s) which are subject to fuzzy (and crisp) constraints.

Case 4: ncf=ncg=nch=0. The optimization problem has only fuzzy objective functions and fuzzy constraints. Consequently the proposed formulation reduces to

$$\max\ a\lambda$$

subject to

$$\lambda \le \mu_{\overline{F}}$$
$$\lambda \le \mu_{\overline{G}_i},\ i = 1,..nfg$$
$$\lambda \le \mu_{\overline{H}_j},\ j = 1,..nfh \tag{39}$$

Case 5: ncf=nfg=nfh=0. The optimization problem has fuzzy objective functions and crisp constraints. A Pareto-optimal solution is obtained by

$$\text{Find } X \in S \text{ which max } a\lambda \text{ such that } \lambda \le \mu_{\overline{F}} \tag{40}$$

Eq. (40) can be restated as

$$\text{Find } X \in S \text{ which max } \mu_{\overline{F}}. \tag{41}$$

Theorem 4: If X^* is an optimum solution to Eq. (41), then X^* is Pareto-optimal.

Proof: See Dhingra and Rao (1995).

Case 6: ncf=0. Case 6 is a generalization of case 4 when crisp constraints are also present in the problem formulation. An optimum solution is obtained by

Find $X \in S$ which max $a\lambda$ subject to

$$\lambda \le \mu_{\overline{F}}$$
$$\lambda \le \mu_{\overline{G}_i},\ i = 1,..nfg$$
$$\lambda \le \mu_{\overline{H}_j},\ j = 1,..nfh \tag{42}$$

Case 7: ncg=nch=0. Case 7 is similar to the original formulation (Eq. 32) except that no crisp constraints are present. An optimum solution is obtained by

$$\max\ a\lambda$$

subject to

$$\lambda \le \mu_{\overline{F}}$$
$$\lambda \le \mu_{\overline{G}_i},\ i = 1,..nfg$$
$$\lambda \le \mu_{\overline{H}_j},\ j = 1,..nfh \tag{43}$$

Case 8: nfg=nfh=0. Case 8 is similar to the original formulation except that the fuzzy constraints are absent. An optimum solution to this problem is obtained by

$$\text{Find } X \in S \text{ which max } B + a\,\lambda \text{ such that } \lambda \le \mu_{\overline{F}} \tag{44}$$

Lemma 8: For $a \geq 0$, if X^* is an optimum solution to Eq. (44), then X^* is weakly Pareto-optimal. If X^* is unique, then it is Pareto-optimal.
Lemma 9: For a > 0, if X^* is an optimum solution to Eq. (44), then X^* is Pareto-optimal.
Theorem 5: If X^* is a optimum solution to Eq. (44), then X^* is Pareto-optimal.
Proof: The proof follows from Thms. 3 and 4.

Case 9: nff=ncf=ncg=nch=0. The optimization problem consists of only fuzzy constraints and no crisp or fuzzy objective functions are specified. An optimum solution to this problem is obtained by

max $a\lambda$

subject to

$$\begin{aligned} &\lambda \leq \mu_{\overline{G}_i},\ i = 1,..nfg \\ &\lambda \leq \mu_{\overline{H}_j},\ j = 1,..nfh \end{aligned} \tag{45}$$

Case 10: nff=ncf=0. The optimization problem has both crisp and fuzzy constraints but no objective functions. An optimum solution is obtained by

Find $X \in S$ which max $a\lambda$ subject to

$$\begin{aligned} &\lambda \leq \mu_{\overline{G}_i},\ i = 1,..nfg \\ &\lambda \leq \mu_{\overline{H}_j},\ j = 1,..nfh \end{aligned} \tag{46}$$

One is most likely to encounter cases 9 and 10 during the preliminary and/or conceptual stages of the design process. During these initial stages a DM rarely has a performance index that he is trying to optimize. All he is attempting to do is come up with a feasible starting design. Cases 9 and 10 are expected to be useful in solving such problems. Once a preliminary design is obtained and judged feasible, it can next be optimized using any of the formulations presented in cases 1 through 8.

It may be noted that for problems with fuzzy constraints, the previous cases make no reference to the Pareto-optimality of the optimum solution. This is because for a crisp MOO problem, Pareto-optimality of a solution is based on the assumption that the solutions can only be classified as being feasible or non-feasible. There is no concept of partial feasibility in crisp MOO theory. However, the solutions to a fuzzy optimization problem differ not only with respect to objective function values, but also with respect to their degree of feasibility. For problems with fuzzy constraints, the concept of fuzzy Pareto-optimality as given by Defn. 5 is needed.

6. NUMERICAL EXAMPLES

The effectiveness of MOO techniques presented in sections 3-5, and the influence of nonlinear membership functions on the overall design process is demonstrated next via an application to two multiobjective problems. The first problem deals with the design of planar high-speed mechanisms for six specified objective functions.

The second example addresses the optimum design of flexible space structures where four objective functions are simultaneously optimized.

6.1 Design Example 1

An integrated approach to the design of planar, high-speed mechanisms is presented next. The proposed multiobjective formulation combines both kinematic and dynamic synthesis aspects of mechanism design. A total of six (one kinematic and five dynamic) criteria are considered. The kinematic criterion deals with the minimization of the difference between the desired motion and the generated motion. The dynamic criteria consist of the minimization of input driving torque, ground bearing forces, and the shaking forces and shaking moments transmitted to the ground link over a cycle. The mathematical formulation of the problem is given next.

A four bar mechanism is to be synthesized to generate a given path with coordinated rotation of the input link. Using Fig. 1, the coordinates of the path traced by the coupler point P are given as

$$\begin{aligned} X_{gi} &= X_{O_A} + r_2 \cos(\theta_{2i} + \alpha) + r_5 \cos(\theta_{3i} + \alpha) - r_6 \sin(\theta_{3i} + \alpha) \\ Y_{gi} &= Y_{O_A} + r_2 \sin(\theta_{2i} + \alpha) + r_5 \sin(\theta_{3i} + \alpha) + r_6 \cos(\theta_{3i} + \alpha) \end{aligned} \tag{47}$$

where $\theta_{2i} = \theta_{2s} + \bar{\theta}_{2i}$. Here X_{O_A} and Y_{O_A} denote the coordinates of the ground pivot O_A, α is the angular orientation of the ground link, r_i (i=1,..,6) are the link lengths, θ_{2s} is the starting position of the input link, and θ_{2i} and θ_{3i} are the angular orientations of links 2 and 3 at the *i*th design position, respectively. Let the desired (prescribed) values of the path coordinates be given as (X_{di}, Y_{di}). The first objective (f_1) considered is a minimization of the structural error over the entire range of motion, and is defined as

$$f_1 = \sum_{i=1}^{N} \varepsilon_i^2 = \sum_{i=1}^{N} \left[(X_{di} - X_{gi})^2 + (Y_{di} - Y_{gi})^2 \right] \tag{48}$$

where *N* denotes the number of points into which the path is divided.

To study the dynamics of the mechanism, the kinetostatic method of analysis (Dhingra and Moskowitz, 1991) is used because we are interested in minimizing the bearing forces and torques resulting from a given mechanism motion. The dynamic analysis is carried out by writing the equations of equilibrium for the free body diagrams of each of the three moving links. This results in a system of nine simultaneous equations with nine unknowns (the *x* and *y* components of the four bearing reactions F_{12}, F_{23}, F_{34} and F_{14}, and the input driving torque T_s). The dynamic analysis is performed at every 5° rotation of the input link. This results in 72 evaluations during each 360° crank rotation. The maximum values of input driving torque (f_2), shaking force (f_3), shaking moment (f_4) and the forces transmitted to the two ground pivots (f_5 and f_6) over a cycle are treated as five objective functions. The maximum values of these quantities, over a complete cycle, are to be minimized.

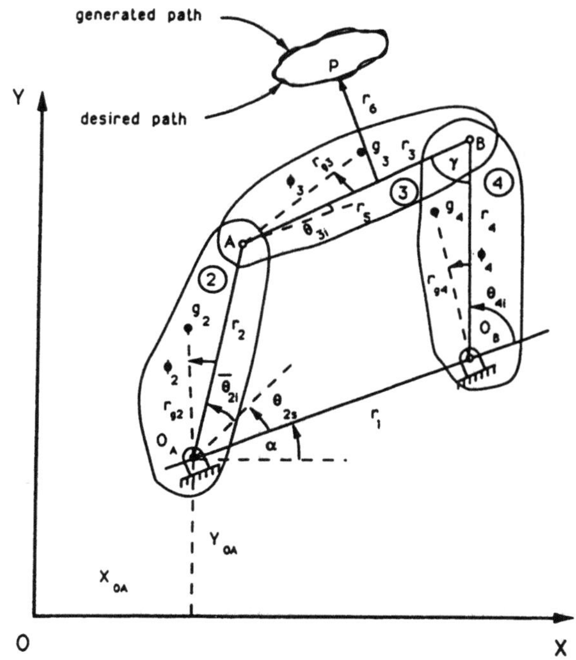

Fig.1 Path generating planar four-bar mechanism.

A minimization of the six (one kinematic and five dynamic) objective functions defined above is achieved by varying the six link lengths (r_1-r_6), three ground pivot O_A parameters, namely X_{O_A}, Y_{O_A} and α (a total of 9 variables), and by adding counterweights to all three moving links. The counterweight radii, thicknesses and orientations are treated as (additional 9) design variables. Thus the optimization problem has a total of 18 design variables.

The problem is subject to the following constraints.

1. The mechanism must satisfy the loop closure equation at each design position. This is achieved by using an equality constraint of the form

$$2r_2r_4\cos(\theta_{2i}-\theta_{4i})-2r_1r_4\cos\theta_{4i}+2r_1r_2\cos\theta_{2i}+r_3^2=r_1^2+r_2^2+r_4^2 \tag{49}$$

at each design position i, $i=1,..,N$.

2. The structural error at each position is constrained to be less than a specified quantity Δ, i.e. $\varepsilon_i \le \Delta$, $i=1,..,N$.

3. A further restriction that insures that the input link be a crank is stated as

$$r_1+r_2<r_3+r_4$$
$$[r_3-r_4]^2<[r_1-r_2]^2 \tag{50}$$

4. The value of transmission angle (γ) over the entire cycle is constrained as

$$\pi / 6 \le \gamma \le (5\pi)/6 \tag{51}$$

5. Lastly, to insure that each solution is free of a branch defect, an inequality constraint relating the sign of γ_0 and γ_j is imposed at each design position j, $j=1,..,N$. In this work, a value of N=10 is used and the coordinates of the prescribed path are

$$X_{di} = 0.4 - \sin 2\pi (t_i - 0.34)$$
$$Y_{di} = 2.0 - 0.9 \sin 2\pi (t_i - 0.5) \tag{52}$$

where $t_i = (i-1)/N$. The coordinated input link orientations are determined using $\bar{\theta}_{2i} = 2\pi\, t_i$. The design problem is subject to a total of 3*N + 4 constraints corresponding to the N design points into which the desired path is divided. For this example, only the objective functions are assumed to be fuzzy.

The single objective optimization problems are solved first, and the optimum values of the objective functions are given in Table 1. Using the results of the single objective optimizations, the membership functions for the six fuzzy goals are constructed. This results in a problem similar to Eq. (12) with 19 variables and 40 constraints. The results obtained by solving this fuzzy optimization problem are summarized in Table 2. Table 2 also presents the results obtained when nonlinear membership functions are used for the six fuzzy objective functions.

Table 1. Objective function values for single objective optimizations.

Obj. Func.	Starting Vector	Min. ε^1	Min. F_{12}^2	Min. F_{14}^3	Min. T_s^4	Min. SF^5	Min. SM^6
$\varepsilon \times 10^2$	55.242	2.63*	65.81	37.64	65.83	17.27	32.4
F_{12}	15.988	15.38	0.0944*	22.791	22.198	0.9837	0.1548
F_{14}	8.5648	7.524	0.1469	0.1051*	0.4149	0.9633	0.1191
T_s	4.6061	3.654	0.0464	0.1413	0.0401*	0.2561	0.0635
SF	18.884	19.34	0.1698	22.882	22.396	0.0561*	0.1648
SM	25.479	23.61	0.1489	41.098	35.933	1.5398	0.1235*

1: Minimize structural error.
2: Minimize maximum value of F_{12}.
3: Minimize maximum value of F_{14}.
4: Minimize maximum value of driving torque.
5: Minimize maximum value of shaking force.
6: Minimize maximum value of shaking moment.
*: Optimal single-objective function values.

It can be seen from Table 3 and Fig. 2 that when all six objectives (f_1-f_6) are considered simultaneously, the dynamic characteristics of the optimum mechanism are significantly better than the linkage for which only the structural error is optimized. The improvement factors for dynamic performance measures range anywhere from 38.6 to 91.8. The structural error has increased by a small amount (1/0.84=1.19) compared to the case when only structural error is minimized. Thus, for this planar mechanism design problem, substantially improved kinematic and dynamic characteristics can be obtained when all the kinematic and dynamic attributes are considered simultaneously.

Table 2. Results obtained with six objectives considered simultaneously.

Obj. Func.	Linear	Hyperb.	Sine	Expon. (k=1)	Quad.	Log	Expon. (k=-1)
$\varepsilon \times 10^2$	3.132	4.121	3.408	3.11	2.847	2.899	3.462
F_{12}	0.1687	0.2634	0.1754	0.1422	0.1882	0.2094	0.1664
F_{14}	0.1949	0.2982	0.2191	0.1735	0.2044	0.2071	0.192
T_s	0.0777	0.0759	0.0941	0.0727	0.0801	0.0737	0.0706
SF	0.2106	0.2865	0.272	0.1986	0.1888	0.1806	0.1954
SM	0.2781	0.4662	0.2649	0.2126	0.2341	0.3029	0.2142
λ	0.989	0.9968	0.9769	0.9865	0.9989	0.9964	0.9939

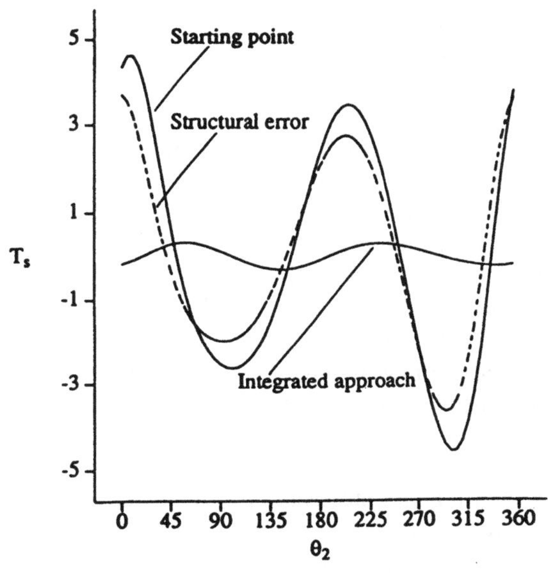

Fig. 2 Driving torque as a function of input link orientation.

The results obtained using the multiobjective formulation also represent a significant improvement over those obtained by Kakatsios and Tricamo (1984). It can be seen from Table 3 that the improvement factors range anywhere from 2.59 (for structural error) to 59.85 (for input driving torque) when comparing the results obtained by Kakatsios and Tricamo to the results obtained using the multiobjective formulation presented herein. Thus prior formulations which have utilized only a single kinematic or dynamic attribute as an objective function, and treated the remaining attributes using inequality constraints, can lead to inferior solutions. This is due to the fact that the conflicting nature of kinematic and dynamic criteria seldom permits selection of a single attribute as an objective function.

Table 3. Comparison of single vs multiple objective optimization results.

Objective	Structural error[1]	Six Objectives[2]	Improv. Factor[3]	Kakatsios Tricamo[4]	Improv. Factor[5]
$\varepsilon \times 10^2$	2.63	3.132	0.84	8.1	2.59
F_{12}	15.38	0.1687	91.17	9.24	54.77
F_{14}	7.524	0.1949	38.61	3.61	18.52
T_s	3.654	0.0777	47.03	4.65	59.85
SF	19.34	0.2106	91.83	9.36	44.44
SM	23.61	0.2781	84.89	12.93	46.49

1: Only structural error (f_1) is minimized.
2: All six objectives considered simultaneously (linear membership functions).
3: Column (4)=Column (2)/Column (3).
4: Results from Table 3, Run 5 in Ref. 16.
5: Column (6)=Column (5)/Column (3).

When nonlinear shapes corresponding to various values of the membership satiation coefficient are used for membership functions of the fuzzy objective functions, the kinematic and dynamic performance measures of the mechanism exhibit similar trends (see Tables 2 and 3). The results obtained using quadratic, exponential ($k < 0$), and logarithmic membership functions are similar, as these three functions model a negative value of m(X). However, the results obtained using these three functions are fairly different from those obtained with membership functions which model constant or increasing marginal membership satiation values. The difference in the results obtained using quadratic, exponential ($k < 0$) and logarithmic functions is attributed to second order effects. Even though these three functions have the same sign for the membership satiation coefficient ($m(X) < 0$), the value of m(X) is constant, decreasing and increasing for these functions. However, the second order effects are not as dominant as the first order effects which govern the sign of membership satiation coefficient. Hence, it is important to accurately assess the nature of the membership functions (e.g., concave, convex, or linear), i.e. the sign of the membership satiation coefficient influences the optimum results significantly.

6.2 Design of Flexible Space Structures

This example presents a multiobjective formulation for the integrated structure-control optimization problem for flexible space structures. The proposed approach also addresses robustness considerations for controller design. The structural weight, controlled system energy, and robustness indices are considered as the objective functions of the integrated structure/control design problem, with cross-sectional areas of the members treated as design variables. The computational procedure is demonstrated through the design of an ACOSS-FOUR space structure.

Large space structures face difficult problems of vibration control. The natural frequency spectrum of these structures is typically quite dense and the passive damping levels are small. To meet the stringent positioning/pointing requirements, it is essential that these structures be actively controlled to achieve the required performance levels. Consequently, considerable interest has recently been generated in the use of active control systems to improve the performance of flexible space structures (Rao et al. 1990, Dhingra et al. 1992b, Dhingra and Lee 1994, 1995).

6.2.1 Equations of Motion

The equations of motion governing the dynamics of a flexible structure are

$$[M]\ddot{U} + [C]\dot{U} + [K]U = [D]F \tag{53}$$

where U is a $n \times 1$ displacement vector, F is a $m \times 1$ control vector, [M], [K], and [C] are $n \times n$ mass, stiffness and damping matrices, respectively, and [D] is a $n \times m$ applied force distribution matrix. Assuming proportional damping and using the transformation $U = [\Phi]\eta$ where $[\Phi]$ is the $n \times n$ modal matrix and η is the vector of modal coordinates, Eq. (53) can be expressed in a state space form as

$$\dot{x} = [A]x + [B]F \tag{54}$$

where $[x] = [[\eta],[\dot{\eta}]]^T$. The plant matrix [A] and the input matrix [B] are

$$[A] = \begin{bmatrix} [0] & [I] \\ [-\omega_i^2] & [-2\varsigma_i\omega_i] \end{bmatrix}, \quad [B] = \begin{bmatrix} [0] \\ [\Phi]^T D \end{bmatrix} \tag{55}$$

where ζ_i and ω_i denote the damping ratio and natural frequency of the *i*th mode. We use the linear quadratic regulator (LQR) theory to design a controller for the structure whose dynamics is governed by Eq. (54). The optimal control F minimizes a quadratic performance index (PI) J_c defined as

$$J_c = \int_0^\infty (x^T[Q]x + F^T[R]F)\,dt \tag{56}$$

where [Q] is a positive semidefinite state weighting matrix, and [R] is a positive definite control weighting matrix. If the system is observable and stablizable (Kwakernaak and Sivan, 1972), there exists a unique optimal feedback control law $F = -[G]x$ with the feedback gain $[G] = [R]^{-1}[B]^T[P]$. Here [P] is a unique positive definite matrix which satisfies the matrix Ricatti equation

$$[A]^T[P] + [P][A] + [Q] - [P][B][R]^{-1}[B]^T[P] = [0] \tag{57}$$

Eq. (54) can now be rewritten as

$$\dot{x} = ([A] - [B][G])x = [A_{cl}]x \tag{58}$$

The closed loop plant given by Eq. (58) is asymptotically stable. The eigenvalues of the closed loop matrix [A_{cl}] are given as $\lambda_i = \alpha_i \pm j\beta_i$ where $j = \sqrt{-1}$ and $|\lambda_i| = \sqrt{\alpha_i^2 + \beta_i^2}$. The closed loop damping ratio ξ_i associated with λ_i is given as $\xi_i = -\alpha_i / |\lambda_i|$, $i=1,..,n$.

6.2.2 Robust Controller Design

The optimum control law used to design a controller for actively controlled structures is frequently based on nominal values of design variables. The parameter variations introduced by the analysis model, and uncertain material and modal properties may adversely effect the stability and performance of the control system. A robust controller is therefore needed which is able to fulfill system specifications satisfactorily even in the presence of modeling errors and parameter uncertainties.

The eigenvalue problem for an exponential solution to Eq. (58) is given as

$$\begin{aligned} &[A_{cl}][\Phi] = [\lambda][\Phi] \\ &[A_{cl}]^T[\Psi] = [\lambda][\Psi] \end{aligned} \tag{59}$$

where $[\Phi]$ and $[\Psi]$ are $2n \times 2n$ matrices containing the right and left hand eigenvectors corresponding to eigenvalues $[\lambda]$ with $\phi_i^H \phi_i = 1$ and $\psi_i^T \phi_j = \delta_{ij}$, i, $j=1,..,2n$. The sensitivity of eigenvalues of the closed loop matrix [A_{cl}] to small perturbations [E] is given by the Bauer-Fike theorem (Golub and van Loan, 1989).

Theorem 6: If μ_i is an eigenvalue of [A_{cl}] + [E] $\in C^{2n \times 2n}$ and $[\Phi]^{-1}[A_{cl}][\Phi] = [\Lambda]$ where $[\Lambda] = diag[\lambda_1, \ldots \lambda_{2n}]$, then $\min|\lambda_i - \mu_j| \le \kappa_p(\Phi)\|E\|_p$, $\lambda_i \in \lambda(A_{cl})$.

Here $\|\bullet\|_p$ denotes the Holder norm with $\kappa_p = \|\Phi^{-1}\|_p \|\Phi\|_p$ and p=2 corresponds to the Euclidean norm. The sensitivity of the eigenvalues of [A_{cl}] to perturbations [E] depends on the condition number of the eigenvector matrix $\kappa_p(\Phi)$. Since $\kappa_p(\Phi)$ establishes the uncertainty region within which the eigenvalues of [A_{cl}] are perturbed, to maximize the stability robustness, $\kappa_p(\Phi)$ must be minimized.

For both the nominal and the perturbed system, good dynamic response can be achieved if the real part of every eigenvalue is constrained to be smaller than a specified value ρ, i.e. $Re(\lambda_i) < -\rho$, $\rho > 0$. The LQR can thus be modified to yield

$$\dot{x} = ([A] + \rho[I])x + [B]F = ([A] + \rho[I] - [B][G])x \tag{60}$$

with $[G] = [R]^{-1}[B]^T[K]$ where [K] is the solution to matrix Ricatti equation with [A] replaced by ([A] + ρ [I]). The closed-loop system given by Eq. (60) has a guaranteed stability margin (ρ).

Discussed next is the effect of changes in system parameters and/or design variables on the performance index J_c. If [P] is the steady state solution to the matrix Ricatti equation, the minimum value of PI J_c is given as

$$J_c^* = x(0)^T[P]x(0) \tag{61}$$

where x(0) is the initial state vector. The performance index of the perturbed system with the same controller (i.e. feedback gains remaining unchanged) can be written as $\bar{J}_c^* = x(0)^T[\bar{P}]x(0)$ where $\bar{P}$ satisfies the Lyapunov's equation

$$[\bar{A}_{cl}]^T[\bar{P}]+[\bar{P}][\bar{A}_{cl}]+[Q]+[G]^T[R][G]=[0] \tag{62}$$

and $[\bar{A}_{cl}]=[\bar{A}]-[\bar{B}][G]$ is the perturbed closed loop system matrix. The relative sensitivity measure used to characterize performance robustness is defined as

$$J_{pr}=(\bar{J}_c^*-J_c^*)/(J_c^*)=x(0)^T([\bar{P}]-[P])x(0)/x(0)^T[P]x(0) \tag{63}$$

Both J_c^* and $\bar{J}_c^*$ depend on the initial state x(0). This result is not very useful since the initial state is not always known. If x(0) is assumed to be a random variable distributed uniformly over the surface of a *2n*-dimensional unit sphere, then a minimization of tr [P] is equivalent to minimizing the expected value of J_c^* over the set of possible states (x(0)). Thus, the performance robustness index given by Eq. (63) can be redefined as $J_{pr}=(tr[\bar{P}]-tr[P])/tr[P]$.

6.2.3 Multiobjective Problem Formulation

The stability robustness index, the minimum value of the PI J_c, the Frobenious norm of control gains, and the structural weight are considered as objective functions of the integrated structure/control design problem. The cross-sectional areas of structural members are treated as design variables.

The first objective function (stability robustness) f_1 measures the sensitivity of the eigenvalues of the closed loop system matrix $[A_{cl}]$ to parameter variations and is defined as $f_1=\kappa_2(\Phi)=\|\Phi^{-1}\|_2\|\Phi\|_2$. The minimum value of the quadratic performance index J_c is chosen as the second objective function to be minimized with $f_2=J_c^*=x(0)^T[P]x(0)$. Since a minimization of tr [P] is equivalent to minimizing the expected value of J_c^* over the set of possible initial states x(0), f_2 is defined as f_2=tr [P]. The weight of the structure given by $f_3=\sum \rho A_i l_i$ with ρ, A_i and l_i denoting the density, cross-sectional area and length of the *i*th member is chosen as the third objective function. Finally, the expected value of the integrand of the quadratic control effort $f_4=E\int_0^\infty x^T[G]^T[R][G]x\,dt=tr[G]^T[R][G]$ is selected as the fourth objective function.

6.2.4 Numerical Results

Consider the ACOSS-FOUR flexible space structure shown in Fig. 3. It models the feed tower in a generic class of large antenna applications. The edges of this tetrahedral truss are 10 units long, and the six bipod legs are each $2\sqrt{2}$ units long. The horizontal base of the tetrahedron formed by nodes 2, 3 and 4 is 2 units above the base plane of reference. The nodal coordinates for the finite element model are given Dhingra et al. (1992b). Four lumped masses of 2 units each are attached at nodes 1 through 4. The sensors and actuators are located in the six bipods, and they are assumed to coincide with each other. The passive damping parameter ζ is assumed to be 0.00001. The Young's modulus and weight density of the structural material are taken to be 1.0 and 0.001 respectively.

The constraints used in the problem formulation include: (i) the closed loop damping ratio ξ_1 must be equal to 0.0145 (h_1); (ii) the imaginary part of the first closed loop eigenvalue should be in the *vicinity* of 1.339 (h_2); (iii) the imaginary part of the second closed loop eigenvalue should be in the *vicinity* of 1.5 (h_3). The italicized words are a source of fuzziness in constraint specifications. Each of the twelve variables representing the cross-sectional of structural members is constrained to lie between 10 and 2000. The state weighting matrix [Q]=10*[I], whereas the control weighting matrix [R]=[I], where [I] is an identity matrix.

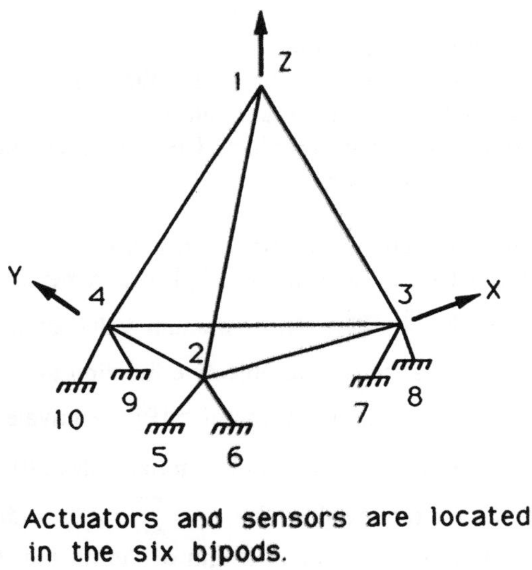

Fig. 3 An ACOSS-FOUR flexible space structure.

The dynamic response of the optimum structure to an initial disturbance, initiated by a unit displacement at node 2 in the *y*-direction at t=0, is also studied by measuring the static displacement associated with the line of sight. Node 1 of the truss represents the antenna feed, and its motion measures the deviation from the line of sight (LOS). The LOS error is given as $LOS = \sqrt{LOS_x^2 + LOS_y^2}$. This should be damped out in order to fall within a certain range in a specified time interval.

Computational procedure

The solution procedure for the integrated structure/control optimization problem consists of the following steps:

1. Assemble the mass, stiffness, and damping matrices for the finite element model of the structure, and determine the natural frequencies and mode shapes.
2. Use the modal data generated in step 1 in conjunction with the sensor/actuator locations to determine the plant matrix [A] and the input matrix [B].
3. Check if the resulting system is observable as well as stablizable.
4. Solve the LQR problem and determine the optimal gain matrix [G].
5. Evaluate all the objective functions and the constraints, and construct membership functions for the fuzzy objective functions and the fuzzy constraints.
6. Update the cross-sectional areas of members using an optimization procedure, and repeat steps 1 through 5 until no further improvement in the objective function(s) is possible.

The single objective optimization problems are solved first, and the optimum values of the objective functions and the cross-sectional areas of the members are given in Tables 4 and 5 respectively. A local minimum is encountered when f_2 is minimized. This is due to the fact that for non-convex problems, standard NLP methods guarantee only a local minimum in the vicinity of the starting point. Using the results of the single objective optimizations, the membership functions for the fuzzy objective functions and the fuzzy constraints are constructed. The membership functions for fuzzy equality constraints are based on a 10% leeway on each side of the desired value. The optimum values of the objective functions and design variables obtained using the fuzzy game theory approach are given in Tables 6 and 7 respectively.

The DM supplied parameter a exercises a considerable bearing on optimum solutions for cases 3,7 and 8. For case 3, parameter a controls a tradeoff between an improvement in the bargaining function B and the degree of constraint violation. The similarity between a, and DM supplied parameter α in the alpha-cut formulation given by Eq. (38) is evident. In the α-cut approach, the decision on a suitable value for α is difficult to make, and the optimum solutions obtained depend on the prescribed value for α. A similar trend was observed for case 3 for various values of a. For cases 7 and 8, a acts as a 'meta-level' weighting parameter between crisp (B) and fuzzy (λ) bargaining functions. Consider the MOO technique, the weighting method, where $F(X) = \sum w_i f_i$ is minimized. The Pareto-optimal solution obtained depends on the weighting factors w_i. A similar trend is observed for cases 7 and 8 where the fuzzy Pareto-optimal solutions obtained depend on the value of a. By varying the value of a, a series of fuzzy Pareto-optimal solutions can be obtained.

Table 4. Objective function values for single objective optimizations.

Objective Function	Starting Point	Min f_1	Min f_2	Min f_3	Min f_4
f_1	81.0866	57.4369	60.7754	59.2866	94.9764
f_2	2.19646	1.69636	1.76267*	1.74322	2.62959
f_3	43.6976	15.1163	19.6058	13.524	43.2807
f_4	173.724	167.056	182.102	138.946	81.9878

* A local optimum is obtained.

Table 5. Design variables for single objective optimizations.

Design Variable	Starting Point	Min f_1	Min f_2	Min f_3	Min f_4
X_1	1000.0	130.403	102.929	166.229	596.224
X_2	1000.0	352.752	539.713	191.084	1444.58
X_3	100.0	143.63	134.179	212.502	94.907
X_4	100.0	200.591	196.951	147.04	104.062
X_5	1000.0	304.394	435.263	216.508	1412.69
X_6	1000.0	199.651	388.766	180.979	104.056
X_7	100.0	108.04	99.706	186.011	409.581
X_8	100.0	117.6	103.65	104.731	406.028
X_9	100.0	109.323	98.274	153.201	417.599
X_{10}	100.0	78.137	90.054	128.91	64.927
X_{11}	100.0	106.545	93.284	195.668	413.73
X_{12}	100.0	117.446	90.455	73.098	308.728

Table 6. Objective function values using cooperative fuzzy games*

Obj.	Case (a)	Case (b)	Case (c)	Case (d)	Case (e)
f_1	62.3528	60.6036	61.8861	60.8793	58.9753
f_2	1.16509	1.28762	1.16971	1.1938	1.30818
f_3	13.5616	13.5238	12.9623	11.912	12.438
f_4	183.362	178.077	181.902	178.76	174.173

*Case (a) f_1-f_4 crisp, h_1-h_3 crisp.
Case (b) f_1-f_4 fuzzy, h_1-h_3 crisp.
Case (c) f_1-f_4 fuzzy, h_1 crisp, h_2 h_3 fuzzy.
Case (d) f_1-f_4 crisp, h_1 crisp, h_2, h_3 fuzzy, a=0.5.
Case (e) f_1, f_2 fuzzy, f_3, f_4 crisp, h_1 crisp, h_2, h_3 fuzzy, a=1.

The results obtained using single and MOO techniques are intuitively appealing. For instance, a minimization of Frobenious norm of control gains (f_4) requires that the structural displacements be small. This, in turn, implies that a stronger structure with large member cross-sectional areas is needed. This, however, results in a heavier design with a poor dynamic response as structural displacements are not damped out quickly enough, thereby leading to a poor value for the quadratic performance index J_c (see Tables 4 and 5). A minimum weight design, on the other hand, requires large control effort to damp out structural displacements.

Table 7. Design variables using cooperative fuzzy games.

Design Variable	Case (a)	Case (b)	Case (c)	Case (d)	Case (e)
X_1	156.147	167.8	149.521	133.065	139.993
X_2	178.687	190.734	174.362	146.726	166.783
X_3	192.718	211.854	171.12	152.03	183.182
X_4	170.244	167.054	153.886	141.653	151.798
X_5	196.499	193.488	195.251	176.218	193.657
X_6	157.746	155.921	144.646	131.965	149.096
X_7	187.972	202.71	228.314	190.027	181.47
X_8	212.141	144.227	209.823	213.821	168.965
X_9	218.383	144.569	201.389	212.221	175.953
X_{10}	171.32	143.96	185.168	190.538	157.899
X_{11}	219.979	234.382	199.326	228.686	169.228
X_{12}	65.374	68.905	62.913	59.073	63.202

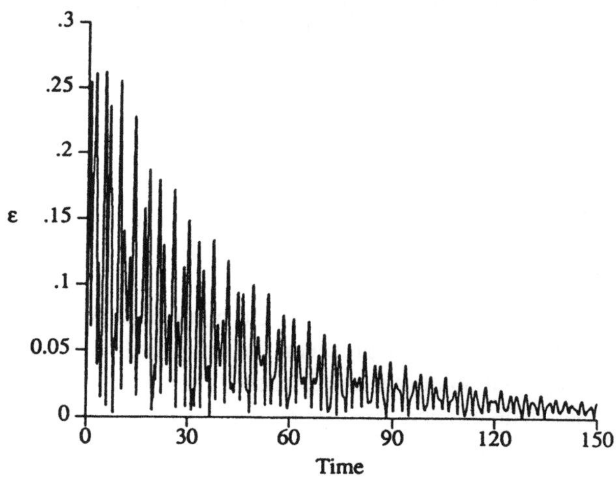

Fig. 4 LOS error versus time for minimum weight design.

The transient response is simulated by finding a solution to Eq. (58) for the period t=0 to t=150 seconds in steps of 0.1 second. Time histories of LOS error when structural weight and Frobenious norm of control gains are minimized are given in Figs. 4 and 5 respectively. It can be seen that the LOS error for the minimum weight design is damped out much more quickly compared to minimum control effort design. The minimum control design is 220% heavier than the minimum weight design, but requires 70% less control energy. Time history of LOS error obtained using the cooperative fuzzy game formulation are given in Figs. 6 and 7. These correspond to cases (a) and (d) in Table 6, respectively. It can be seen from Table 6 and Figs. 6 and 7, that the optimum design corresponding to case (d) is 15% lighter than that corresponding to case (a), has a better stability robustness index and requires less control effort. However, the LOS error corresponding to case (d) is damped out at a rate slower than that for case (a). Further, the optimum designs show considerable improvement over the nominal design which, underscores the importance of integrating the structure and control system design process.

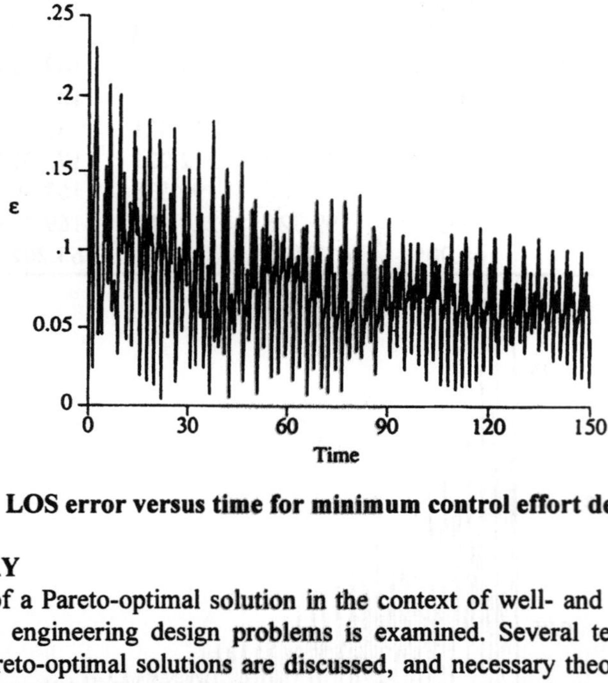

Fig. 5 LOS error versus time for minimum control effort design.

7. SUMMARY

The concept of a Pareto-optimal solution in the context of well- and ill-structured multiobjective engineering design problems is examined. Several techniques for generating Pareto-optimal solutions are discussed, and necessary theorems for optimality are outlined. The use and effectiveness of MOO techniques is demonstrated via an application to two design problems.

The set of Pareto-optimal solutions and the compromise solution generated by each crisp MOO technique are, in general, different from each other. This is due to the fact that the preference structure underlying each approach is different. Methods such as the global criterion formulation do not require any preference information from the DM and can yield a compromise solution for which a particular objective function value may be completely unacceptable to the DM. Utility function, bounded objective, goal programming, goal attainment, and lexicographic methods require an apriori articulation of preference information from the DM. The main advantage of utility function methods is that if the utility function is correctly assessed, it will insure a most satisfactory solution to the DM. However even for a simple problem, the assessment of utility function is difficult. The bounded objective, goal programming and goal attainment methods are computationally efficient, but yield solutions which are fairly sensitive to the goal vector prescribed by the DM. Therefore, one should exercise caution when selecting a particular technique to solve crisp MOO problems.

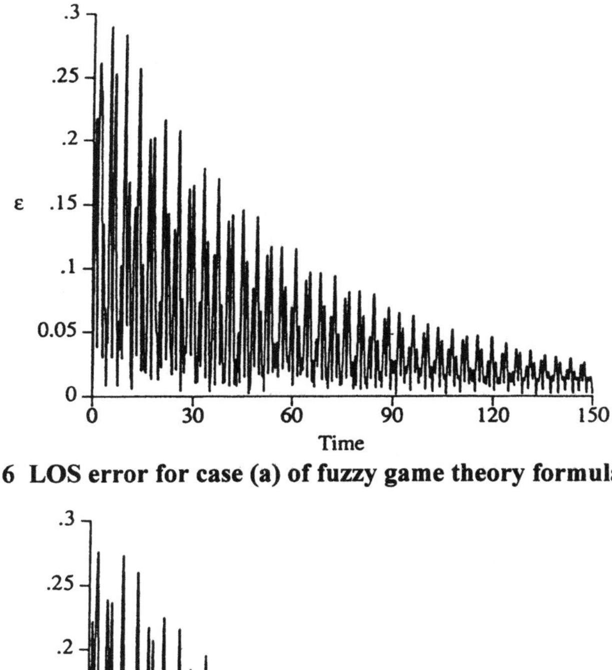

Fig. 6 LOS error for case (a) of fuzzy game theory formulation.

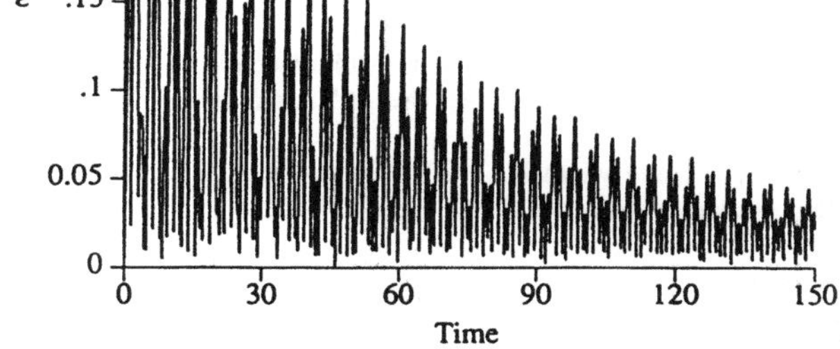

Fig. 7 LOS error for case (d) of fuzzy game theory formulation.

Ill-structured optimization problems where the problem data, objective functions, and constraints are not known precisely, are solved using fuzzy set theory. Three techniques for solving single and multiobjective fuzzy optimization problems are presented. It is shown that the fuzzy formulations yield Pareto-optimal and fuzzy Pareto-optimal solutions. A commonly used assumption in fuzzy optimization applications involving the use of linear membership functions is also examined to determine its impact on the overall design process. It is seen that the final designs are strongly influenced by the sign of membership satiation coefficient. Second order effects which govern whether the satiation coefficient is increasing, decreasing, or constant over the range of interest were found not influence the results significantly.

The applicability of cooperative game theory in an engineering design process is also examined. It is demonstrated how game theory may be used as a tool for solving MOO problems. The concepts in cooperative game theory and fuzzy set theory are combined to yield an optimization approach referred to herein as cooperative fuzzy games. Cooperative fuzzy game theory can be used to solve not only well- and ill-structured, single and MOO problems, but also design problems with no objective functions. A formulation capable of solving problems with partly crisp and partly fuzzy objective functions, as well as partly crisp and partly fuzzy constraints is presented. The fuzzy optimization techniques discussed herein should be more generically useful in solving a multitude of design problems in which deterministic or probabilistic techniques are difficult or impossible to use due to the inherent qualitative, imprecise, or subjective nature of the problem formulation.

8. Acknowledgments
The author would like to thank his former advisor Prof. S.S. Rao for introducing him to the fascinating field of multiobjective optimization and fuzzy set theory. Portions of this work were completed during author's stay at Purdue University and the Corporate Research and Development Center, General Electric Company.

REFERENCES

1. Bellman, R.E., and Zadeh, L.A., 1970, "Decision Making in a Fuzzy Environment," *Management Science*, Vol. 17, pp. B141-164.

2. Dhingra, A.K., Rao, S.S., and Miura, H., 1990a, "Multiobjective Decision Making in a Fuzzy Environment with Applications to Helicopter Design," *AIAA Journal*, Vol. 28, No. 4, pp. 703-710.

3. Dhingra, A.K., Rao, S.S., and Miura, H., 1990b, "Pareto-Optimal Solutions in Helicopter Design Problems," *Engineering Optimization*, Vol. 15, pp. 211-231.

4. Dhingra, A.K., and Rao, S.S., 1991, "An Integrated Kinematic-Kinetostatic Approach to Optimal Design of Planar Mechanisms Using Fuzzy Theories," Trans. of ASME: *Journal of Mechanical Design*, Vol. 113, No. 3, pp. 306-311.

5. Dhingra, A.K., and Moskowitz, H., 1991, "Application of Fuzzy Theories to Multiple Objective Decision Making in Systems Design," *European Jl of Operational Research*, Vol. 53, pp. 348-361.

6. Dhingra, A.K., Rao, S.S., and Kumar, V., 1992a, "Nonlinear Membership Functions in Multiobjective Fuzzy Optimization of Mechanical and Structural Systems," *AIAA Journal*, Vol. 30, No. 1, pp. 251-260.

7. Dhingra, A.K., Rao, S.S., and Kumar, V., 1992b, "Robust Design of Actively Controlled Structures Using Cooperative Fuzzy Games," *Engineering Optimization*, Vol. 20, pp. 81-109.

8. Dhingra, A.K., and Lee, B.H., 1994, "Optimal Placement of Actuators in Actively Controlled Structures," *Engineering Optimization*, Vol. 23, pp. 99-118.

9. Dhingra, A.K., and Rao, S.S., 1995, "A Cooperative Fuzzy Game Theoretic Approach to Multiple Objective Design Optimization," *European Journal of Operational Research*, (in press).

10. Dhingra, A.K., and Lee, B.H., 1995, "Multiobjective Design of Actively Controlled Structures Using a Hybrid Optimization Method," *Intl. Jl. for Numerical Methods in Engineering* (in press).

11. Dyer, J.S., and Sarin, R.K., 1982, "Relative Risk Aversion," *Management Science*, Vol. 28, No. 8, pp. 875-886.

12. Evans, G.W., 1984, "An Overview of Techniques for Solving Multiobjective Mathematical Programs," *Management Science*, Vol. 30, No. 11, pp. 1268-1282.

13. Feng, Y., 1983, "A Method Using the Fuzzy Mathematics to Solve the Vectormaximum Problem," *Fuzzy Sets and Systems*, Vol. 9, pp. 129-136.

14. Geoffrion, A.M., 1968, "Proper Efficiency and the Theory of Vector Maximization," *Journal of Mathematical Analysis and Applications*, Vol. 22, pp. 618-630.

15. Golub, G.H., and van Loan, C.F., 1989, *Matrix Computations*, The Johns Hopkins University Press, Baltimore, Maryland.

16. Kakatsios and Tricamo, 1984, "Simultaneous Optimization of Kinematic and Dynamic Characteristics in the Design of Planar Mechanisms," *ASME Paper No. 84-DET-43.*

17. Kim, K.J., Moskowitz, H., Dhingra, A.K., and Evans, G., 1991, "Fuzzy Multicriteria Methodologies for Quality Function Deployment," *Center for the Management and Manufacturing Enterprises*, Purdue University, Report No. 91-10-01.

18. Koski, J., 1993, "Multicriteria Optimization in Structural Design: State of the Art," in *Advances in Design Automation*, DE-Vol. 65-1, ASME, pp. 621-629.

19. Kuhn, H.W. and Tucker, A.W., 1951, "Nonlinear Programming," in *Proc. of the Second Berkeley Symposium on Mathematical Statistics and Probability*, J. Neyman (Ed.), pp. 481-492.

20. Kumar, V., Dhingra, A.K., and Cakal, H., 1990, "A Unified Approach for Multiobjective Design Optimization," Proc. of *3rd Airforce/NASA Symp on Multidisciplinary Analysis and Optimization.*

21. Kwakernaak, H., and Sivan, R., 1972, *Linear Optimal Control Systems*, Wiley, New York.

22. Leitmann, G., 1975, "Cooperative and Non-Cooperative Differential Games," in *Multicriteria Decision Making*, G. Leitmann and A. Marzollo (eds.), Springer Verlag, New York, pp. 7-31.

23. Luhandjula, M.K., 1989, "Fuzzy Optimization: An Appraisal," *Fuzzy Sets and systems*, Vol. 30, pp. 257-282.

24. Nash, J.F., 1950, "The Bargaining Problem," *Econometrica*, Vol. 18, pp. 155-162.

25. Rao, S.S., 1987a, "Description and Optimum Design of Fuzzy Mechanical Systems," *Journal of Mechanisms, Transmissions, and Automation in Design*, Vol. 109, No. 1, pp. 126-132.

26. Rao, S.S., 1987b, "Multi-Objective Optimization of Fuzzy Structural Systems," *International Journal for Numerical Methods in Engineering*, Vol. 24, pp. 1157-1171.

27. Rao, S.S., Pan, T.S., and Venkayya, V.B., 1990, "Modeling and Control of Flexible Structures: A Survey," *Applied Mechanics Reviews*, Vol. 43, No. 5, pp. 99-117.

28. Soland, R.M., 1979, "Multicriteria Optimization: A General Characterization of Efficient Solutions," *Decision Sciences*, Vol. 10, No. 1, pp. 26-38.

29. Stadler, W., 1984, "Multicriteria Optimization in Mechanics (A Survey)," *Applied Mechanics Reviews*, Vol. 37, No. 3, pp. 277-286.

30. Stadler, W. and Dauer, J., 1992, "Multicriteria Optimization in Engineering: A Tutorial and Survey," in *Structural Optimization: Status and Promise*, M. Kamat (ed.), AIAA, pp. 209-249.

31. Wang, G.Y., and Wang, W.Q., 1985, "Fuzzy Optimum Design of Structures," *Engineering Optimization*, Vol. 8, pp. 291-300.

32. Zadeh, L.A., 1963, "Optimality and Non-Scalar Valued Performance Criteria," *IEEE Transactions on Automatic Control*, Vol. AC-8, pp. 59-60.

33. Zadeh, L.A., 1965, "Fuzzy Sets," *Information and Control*, Vol. 8, pp. 338-353.

34. Zadeh, L.A., 1973, "Outline of a New Approach to the Analysis of Complex Systems and Decision Process," *IEEE Transactions on Systems, Man, and Cybernetics*, Vol. SMC-3, pp. 28-44.

35. Zimmermann, H.J., 1978, "Fuzzy Programming and Linear Programming with Several Objective Functions," *Fuzzy Sets and Systems*, Vol. 1, pp. 45-55.

36. Zimmermann, H.J., 1986, "Fuzzy Set Theory and Mathematical Programming," in *Fuzzy Set Theory and Applications*, A. Jones et al. (eds) , D. Reidel Publishers, Dordrecht, pp. 99-114.

37. Zimmermann, H.J., 1991, *Fuzzy Set Theory and its Applications*, 2nd edn, Kluwer Academic Publishers, Boston.

38. Zimmermann, H.J., 1992, "Methods and Applications of Fuzzy Mathematical Programming," in *An Introduction to Fuzzy Logic Applications in Intelligent Systems*, R. Yager and L. Zadeh (eds.), Kluwer Academic Publishers, Boston, pp. 97-120.

3

INTELLIGENT SYSTEMS FOR CONFIGURATION PROBLEMS

Prof. Dr. Hans-Jürgen Sebastian

Rheinisch-Westfälische Technische Hochschule Aachen
Institut für Wirtschaftswissenschaften
Operations Research
52056 Aachen, Germany

1 Intelligent Decision Support Systems for Design and Configuration Tasks

1.1 Intelligent Decision Support Systems

The complexity of the strategic decision-making environment is the main reason Intelligent Decision Support Systems (IDSS, also called Active DSS - ADSS) are needed. Complexity arises for several reasons:

- the information and knowledge for the decisions is incomplete, uncertain or imprecise or even inconsistent, but
- the information overload is increasing
- there are multiple, often conflicting goals and multiple constraint types
- there are time constraints for the decisions in a changing environment and
- there is a tendency towards group decision making.

Intelligent DSS are needed in many different type problem domains, e.g. in Engineering, Management, Medicine and Environmental Change (see Fig. 1). The following methods, technologies and techniques are relevant to approach the aim "Intelligent DSS" (Fig. 2).

If an IDSS will be developed for a problem, some of the technologies mentioned in Fig. 2 should be combined to approach a system which shows intelligent abilities. Consequently IDSS are typically to be hybrid systems.

In this chapter "Synthesis-type tasks" or "composition type problems" will be consi-

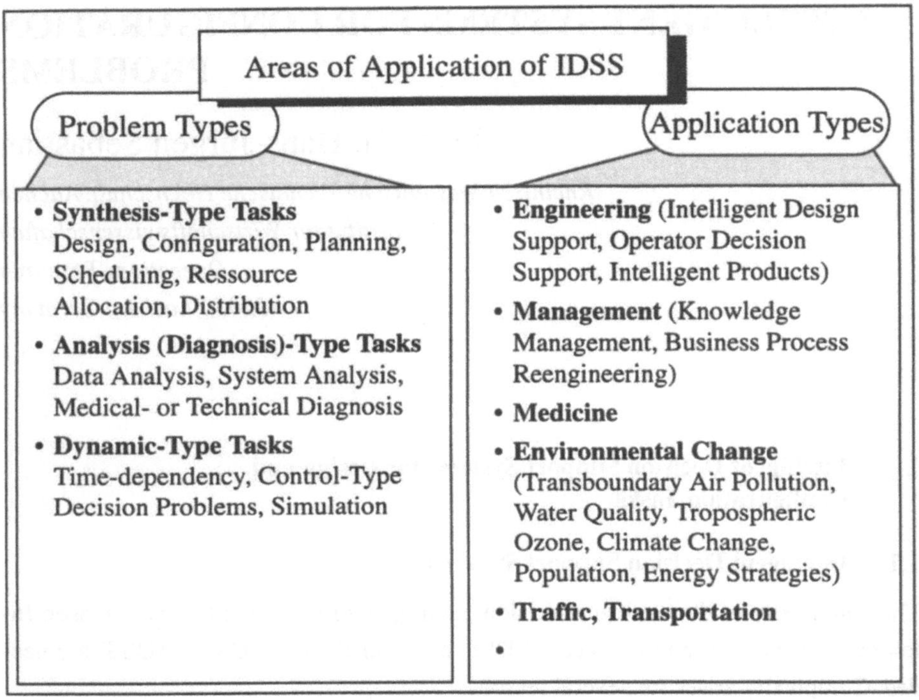

Figure 1: Areas of Application of IDSS

dered. Engineering design and configuration problems as well as planning and scheduling problems belong to that class. Composition type problems are characterized by specification of abstract entities (generic objects), by more and more precise entities and by composition of more complex entities (called systems) of components [4], [25], [39], [51].

The focus of this chapter is on Intelligent Systems for Design and Configuration Problems with applications in the Engineering field. However, to illustrate the wide range of applications, we will consider also an OR-example (logistic layout) and an example of supporting the modelling process in the environmental research field.

The hybrid type systems we are proposing are characterized by integration of some of the technologies mentioned in Fig. 2:

- Knowledge-based Systems
- Fuzzy Logic
- Multi-Criteria Decision Making (MCDM)

Advanced OR technologies such as

- Multi Criteria Decision Making (MCDM)
- Combinatorial Programming
- Natur-Analogous Optimization Strategies (Neural Nets, Evolutionary Algorithms, Simulated Annealing, Tabu Search)

combined with:

Artificial Intelligence: Knowledge-based Systems; Modeling Languages, Approximate Reasoning, Case-Based-Reasoning, Constraint Propagation and Satisfaction

Fuzzy Logic and Fuzzy Technologies: Fuzzy Decision Making, Fuzzy Optimization, Linguistic Approximation, Fuzzy Data Analysis, Fuzzy Data Banks and Query Languages, Fuzzy Hierarchies

Soft Computing: Integration of Fuzzy Logic - Neural Nets - Evolutionary Computing

Figure 2: Relevant Methods for IDSS

Because modelling of imprecision by Fuzzy Logic is the focus of this book, knowledge-based and optimization-based design will first be presented so that the integration of Fuzziness into those techniques will be clear later.

1.2 Innovative Design And Configuration Tasks

First we want to give a definition of design tasks. Tong [49] defines design as follows:

> "Design is the process of constructing a description of an artifact that satisfies a (possibly informal) functional specification, meets certain performance criteria and resource limitations, is realizable in a given target technology, and satisfies criteria such as simplicity, testability, manufacturability, reusability etc.; the design process itself may also be subject to certain restrictions such as time, human power, cost etc."

In addition to this general definition we understand engineering design to be design of "*physical artifacts or physical processes of various kinds*". Generation of structure from a functional description of requirements is central to this kind of design.

Design tasks are a subclass of synthesis tasks. In contrast to classification tasks they are usually solved by stepwise refinement. One subclass of synthesis tasks - configuration - is well investigated. In [34] it is described as follows:

"Configuration - as well as planning - is a synthesis task. A configuration task is understood as the assembling of a target system (a configuration) with predefined architecture and functional behavior from a set of components (objects), (Fig. 3).

Configuration Problem

Find a consistent (or optimal) configuration of a car using predefined alternatives for the car or for the components of a car according to the requirements of a customer

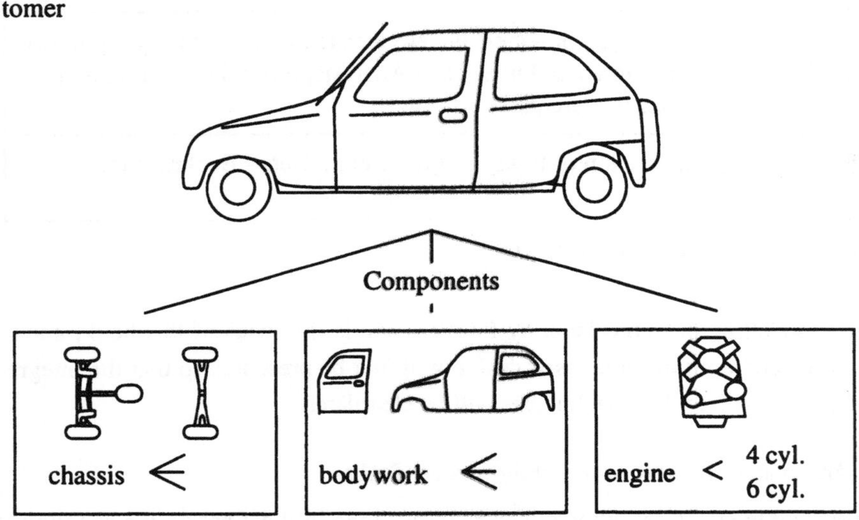

Figure 3: Simple illustrative example of a configuration problem

Using this definition we understand configuration to be an innovative type of design, because the objects and their properties need not to be explicitly known but might be defined implicitly only by a set of constraints. Of course, routine design, where for each object there is a finite set of explicitly given alternatives known in advance is a special case of configuration.

The systems we are elaborating are not creative design systems, but to some extent they are assisting the creativity of a human designer considerably.

The application area of those kind of configuration problems is large. For example, one might easily find configuration problems for

- Computer, Computer-Nets
- Communication or Automation Systems
- Vehicles (Car, Space Launch Systems ...)
- Special Machines, Machine-Systems, Plants
- Architecture (House Design, ...)
- Chemical Engineering (Molecules)
- Layout Problems (Cabin Layout of an Airplane, Layout of a Factory, Logistics Layout)
- Financial Engineering
- Models or Software Systems

in [6], [18], [22], [23], [24], [25], [49]

Most of the applications are from (or closely connected to) Engineering fields. But one can also consider the "design of contracts" (financial engineering) or the design of a mathematical model or a software system as an innovative type configuration task as explained before.

In this chapter we will consider 5 applications from different (mostly engineering) areas, where Fuzziness should be involved.

1.3 An Architecture of an Intelligent DSS for Design and Configuration Tasks

Before we go deeper into the details of our approach to innovative design problems we give as an outline an architecture of an IDSS for supporting design or configuration. Of course, such an architecture gives a very general view of the system we will describe in detail later in Section 4. However, it shows the main components and gives a feeling how the design process can proceed with the help of such a system (Fig. 4).

The knowledge about the domain, where the target system is located is represented in the "Domain-Knowledge-Base" by hierarchies of objects and constraints. A second knowledge-base contains the knowledge about MCDM and/or Optimization Models - the so called "Model-Base". There is a third knowledge-base which represents

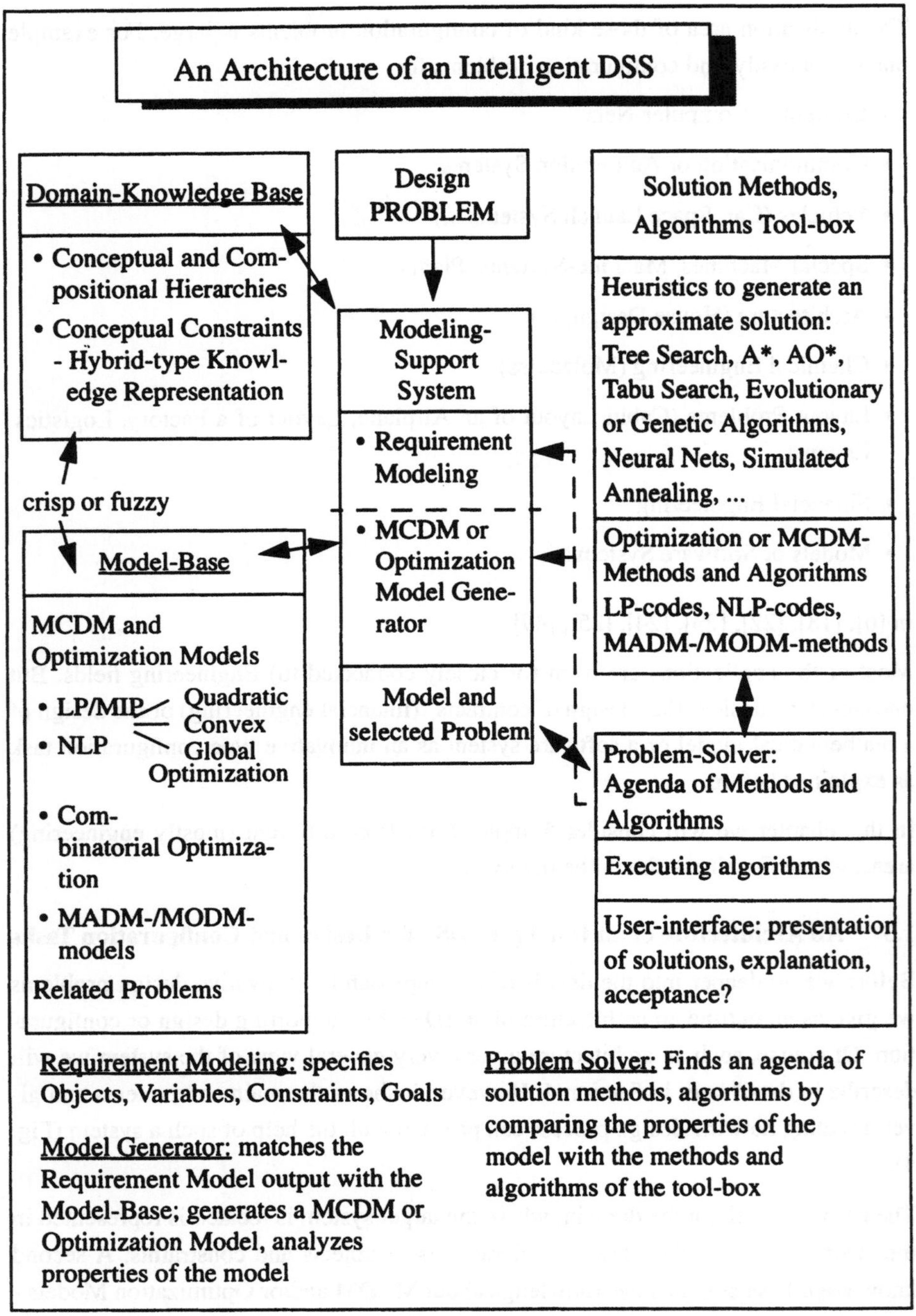

Figure 4: An Architecture of an IDSS

solution methods and algorithms to solve the MCDM or Optimization problems generated by the user and/or system. These three knowledge-bases consist of crisp and imprecise (fuzzy) knowledge.

We find Fuzzy Logic at several places in this architecture, e.g. in the

• Domain Knowledge-Base:	Fuzzy Frames/Object Hierarchies Fuzzy Constraints
• Model-Base	Fuzzy MADM and Fuzzy MODM models and problems
• Solution Methods, Algorithms Toolbox	Fuzzy MADM/MODM methods and algorithms dealing with Fuzziness

The kernel of the system is the Modelling Support System. The Requirement Model assists the user to add his/her requirements to the domain knowledge base starting with the task description. Communicating with both the Model-Base and the Requirement-Model the "MCDM or Optimization Model Generator" generates a model/problem and analyses its properties automatically. Finally the problem solver generates an agenda of algorithms that are most promising to solve the problem successfully by matching the properties of the model/problem with the knowledge about applicability of algorithms to models/problems. After executing the top algorithms of the agenda the problem is, hopefully, solved and the results are given back to the knowledge base and to the Algorithms Toolbox as well to learn about applicability of algorithms. With this architecture and the related ideas of the design process in mind, we will now consider the components of the architecture more in detail.

2 Modelling Design and Configuration Problems Including Fuzziness

The approach described in this section was partly published in K. Hirota's book [23] and in [46]. In [23], together with A. Günter and M. Kopisch we also described KONWERK and two real world applications as well - the passenger aircraft cabin layout problem and the preliminary design and configuration problem of future space launch systems (see also sections 3, 4 and 5).

2.1 The Knowledge Model

We start by describing the domain-knowledge base. There are several ways published in the literature to do this. Most of them are using hierarchies of objects to describe specialization and component structure a target system, using different kinds of notation, e.g. directed graphs with marked edges.

To introduce the related concepts we use simple illustrative examples in Section 2, but in Section 3 we use real world applications.

Fig. 5 shows an example of a conceptual hierarchy to model a "car"-domain. The "has-parts with number specification" and the is-a inverse relation are modelling the "component" and the "specialization" structure of generic objects, also called concepts or classes in the following.

To model the knowledge about a Car-domain more precisely the objects (nodes) must be characterized by attributes and the relations between attributes (constraints) must be represented. Before doing this, however, we give a more systematic approach of a conceptual hierarchy first.

For the representation of the domain knowledge we use hierarchies to model

- the world of systems under consideration (is-a-hierarchy)
- the world of elements (components) of these systems (is-a-hierarchy) and
- the relations between systems and elements (part-of-hierarchy).

The part-of-hierarchies may be considered as multilevel hierarchies, because a component may also be considered as a system consisting of components of the second level.

It is possible to use only one "conceptual hierarchy" to represent all the hierarchies mentioned above. In Fig. 6 such a conceptual hierarchy is introduced where:

S	- denotes the most general system (related to the world of systems),
S_i	- are the system variants (conceptual level),
$OS_{i,k}$	- are system instances (individuals),
E_j	- denotes the components of S_j (conceptual level),
$E_{j,r}$	- are component variants (conceptual level) and
$OE_{j,r,e}$	- are the component instances (individuals).

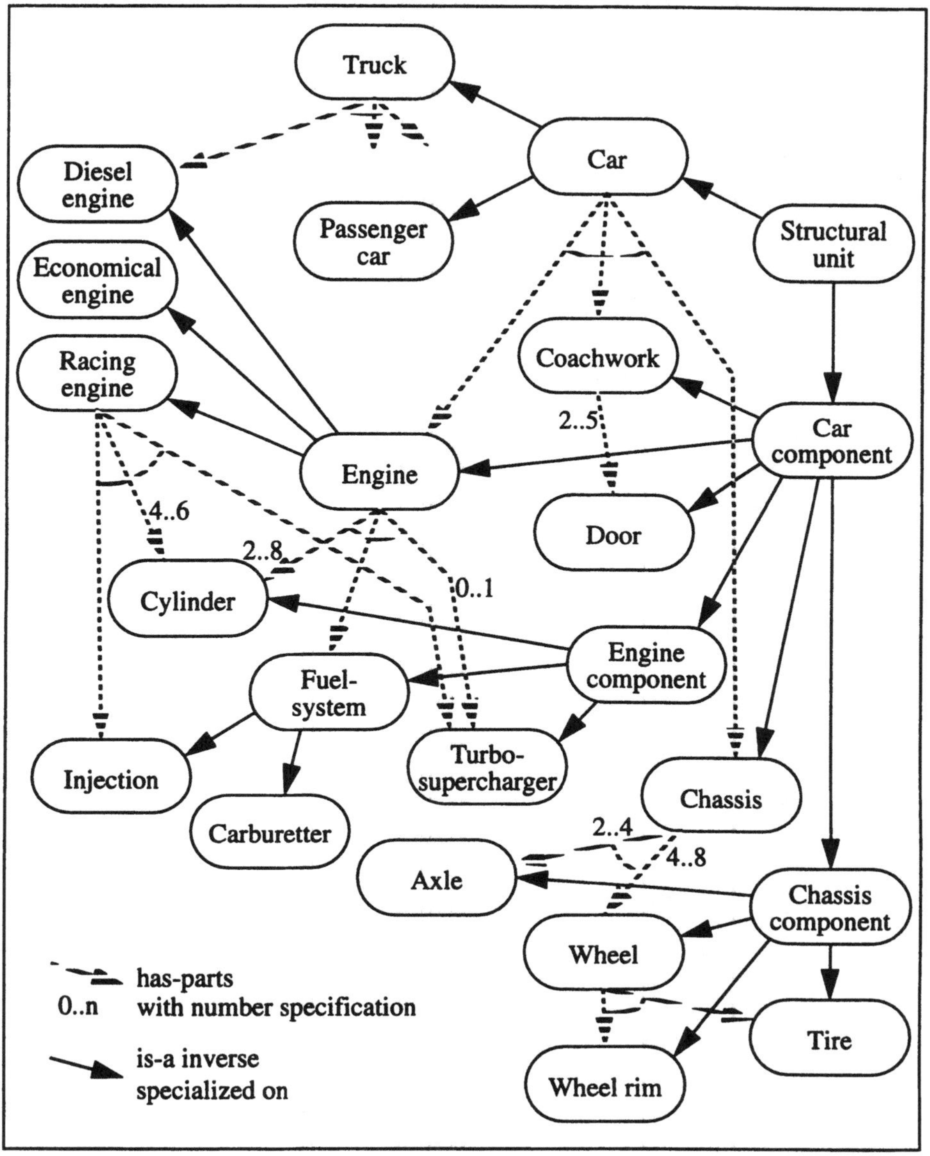

Figure 5: Conceptual Hierarchy - Car

In Fig. 6, for the representation of the basic domain knowledge, we use abstract entities (concepts, generic objects, classes) S, S_i, E_j, $E_{j,r}$ and instances (individuals, specified objects) as well. These objects are connected by relations of the types:

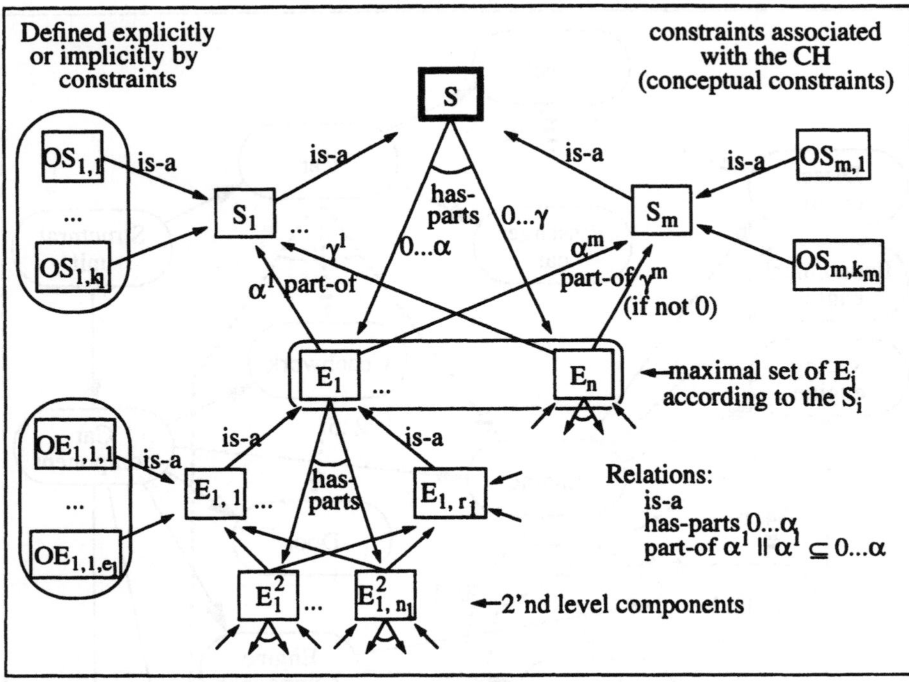

Figure 6: Conceptual Hierarchy - a general view

is-a (specialization) (1-1 relation)

has-parts (composition) (1-n relation) with arc-quantification $0..\alpha$

part-of (1-1 relation) with arc-quantification α^1

Fig. 7 shows an illustrative example of a conceptual hierarchy introduced in Fig. 6.

Design, or as a special case configuration selection, means specification of abstract entities by more precise entities or composition of the precise entities (called specified objects or object classes) to build a system.

The specification process starts with the abstract entities (components) and runs over several stages of object classes, terminating with precise entities (object-instances or individual-instances).

For a formal description we denote by Y either a system S, a component E, a specialized concept or an individual O.

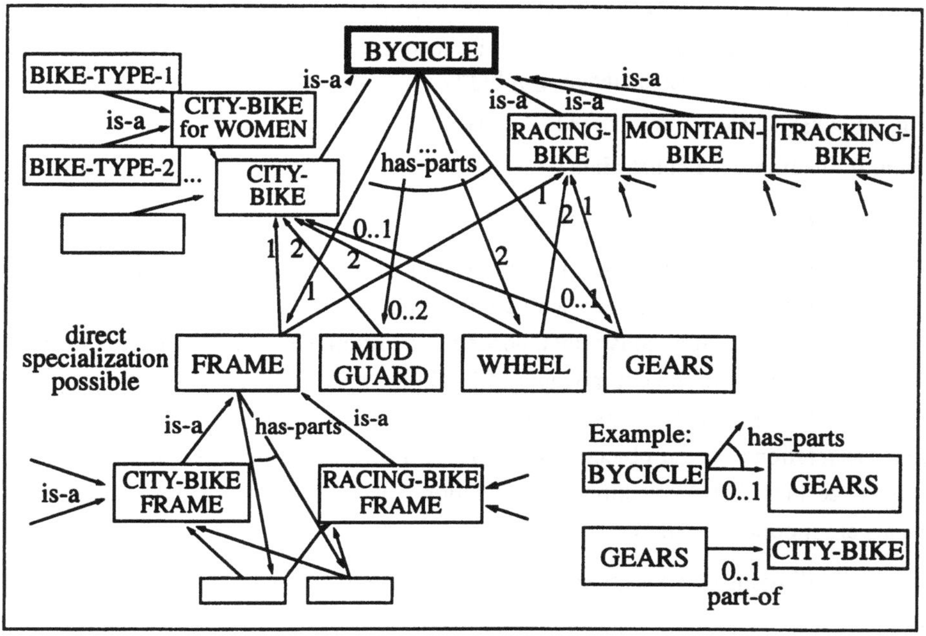

Figure 7: Conceptual Hierarchy - an example

Then, Y is described by

- a name N(Y)
- a vector y = (p, e) with

  p - vector of numerical parameters

  e - vector of linguistic parameters

  $p \in \mathbb{P}, e \in \mathbb{E}$ ($\mathbb{P}$, $\mathbb{E}$ given sets)

  $Y = (N(Y), y)$

The components of p and e are either crisp or fuzzy values.

If E_i denotes a component belonging to the set $\mathbb{E} = \{E_1, E_2, ..., E_n\}$ of elements of S, then $\tilde{O}_i$ (S) denotes the set of all object-instances of E_i.

- Routine design means, that there is a finite number of predefined object-instances $O_i = O_{i,\alpha}$ for E_i, $\alpha = 1, 2, ..., \alpha_i$ which are explicitly known to the configuration system.

  ($O_{i,\alpha} = N(Y)$, Y is an object from $\tilde{O}_i$ (S).)

  $\tilde{O}_i = (O_{i,1}, O_{i,2}, ..., O_{i,\alpha_i})$

- Innovative design (special case) means that not all object-instances are explicitly predefined and known. But there are finite sets of well-known constraints, which describe feasible object-instances implicitly.

  $g_j : \mathbb{P} \times \mathbb{E} \rightarrow \mathbb{R}^1, \qquad j=1, 2, \dots j_0.$

Then, for example the constraints are

$$g_j(y) \leq 0 \qquad \wedge \qquad y \in \mathbb{P} \times \mathbb{E}.$$

The implementation of the specification process might be realized by a FRAME-representation language combined with a Constraint-language (for example PLAKON [6]). There are structural constraints, needed in addition, because we have to describe the relations between the components of S. On an abstract level a structure graph G (S) = (E(S), R(S)) models these relations and the object-structure graph GO(S) models the relations on the object-instance level. For more precise definition of these graphs see [4], [39]. In Section 4 and Section 5 we show in detail how such constraints are modelled in our real world examples.

Now we will focus on conceptual hierarchies with imprecise attributes.

The conceptual hierarchy might contain objects with imprecise attributes (called imprecise objects). For instance, an attribute of an object might be defined as a linguistic variable in which the terms are characterized by the corresponding fuzzy sets. Also, an attribute might be a fuzzy number. Objects, consequently, are described by a few crisp or fuzzy attributes. Therefore, in such hierarchies with imprecise objects, one has to define carefully what specialization and generalization means. In particular, multiple inheritance and inference in fuzzy conceptual hierarchies have to be defined and implemented. The basic problems, as well as implementation oriented aspects connected with these hierarchies, are solved in KONWERK (discussed in Section 4).

In the following, we again consider a simple illustrative example to introduce the ideas:

Referring to a frame like language we use in our implementation, we denote the objects in the following as frames.

In Fig. 8 FRAME TUBE is generated for the generic object tube (also called class tube) by introducing the attributes (slots) by a name, a type and a characterization of precision or imprecision of the attribute. The subconcept (subclass) TUBE-curved requires the definition of an additional attribute CURVATURE. Fig. 9 shows the specialization of a class to a subclass (e.g. an instance or individual). This shows that during the design-process (e.g. specialization process) imprecision will be reduced or even eliminated.

Name of Object	Attributes		
TUBE	Name	Type (linguistic/numeric)	Precision (crisp/fuzzy)
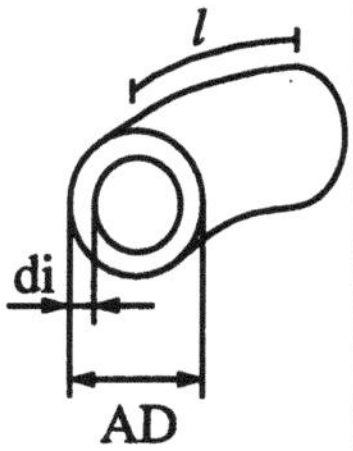	1) FORM	L_1 = {straight, curved}	crisp
	2) MATERIAL	L_2 = {steel, lead, plastics, ceramics, ..}	crisp
	3) THICKNESS	N_3 = $[di_{Min}, di_{Max}]$ real	crisp
	4) DIAMETER	L_3, L_4 = {thin, medium, thick}	fuzzy
		N_4 = $[AD_{Min}, AD_{Max}]$ real	crisp
	5) LENGTH	N_5 = $[l_{Min}, l_{Max}]$	crisp
		L_5 = {very-short, short, medium, long, very-long}	fuzzy

if curved, then additional attribute curvature k (radius)
- crisp
- fuzzy → weak-curvature, medium-curvature, strong-curvature

↑ value sets
in fuzzy cases: membership-functions added

Figure 8: Conceptual Hierarchies (Hierarchies of Objects described by Attributes)

Crisp objects in the sense of individuals belong not only to one, but in general to several imprecise classes (concepts). The membership-degree is dependent mainly on the interpretation of the fuzzy-"and", Fig. 10 and 11 show a very simple example. In Fig. 10 and 11 the membership-degrees 0.4 or 0.3 are computed by Min-Operator application using the membership-functions of the objects which are on the respective path from the considered object node (complete-specified imprecise class) to the root-node of the tree.

$$\mu_{O^*_{4,1,4}}\text{(Object-1)} = \mathrm{Min}\,(\mu_4\,(K),\ \mu_1^*\,(di)\ ,\ \bar{\mu}_4\,(l)\ ) = \mathrm{Min}\,(0.4, 0.5, 0.3) = \underline{0.3}$$

$$\mu_{O^*_{4,1,3}}\text{(Object-1)} = \mathrm{Min}\,(\mu_4\,(K),\ \mu_1^*\,(di)\ ,\ \bar{\mu}_3\,(l)\ ) = \mathrm{Min}\,(0.4, 0.5, 0.6) = \underline{0.4}$$

The membership-values used in these formulas one gets from the respective membership-functions which are available as slot-descriptions of the respective frames.

For a more detailed description of the Fuzzy Frame Hierarchies see [42].

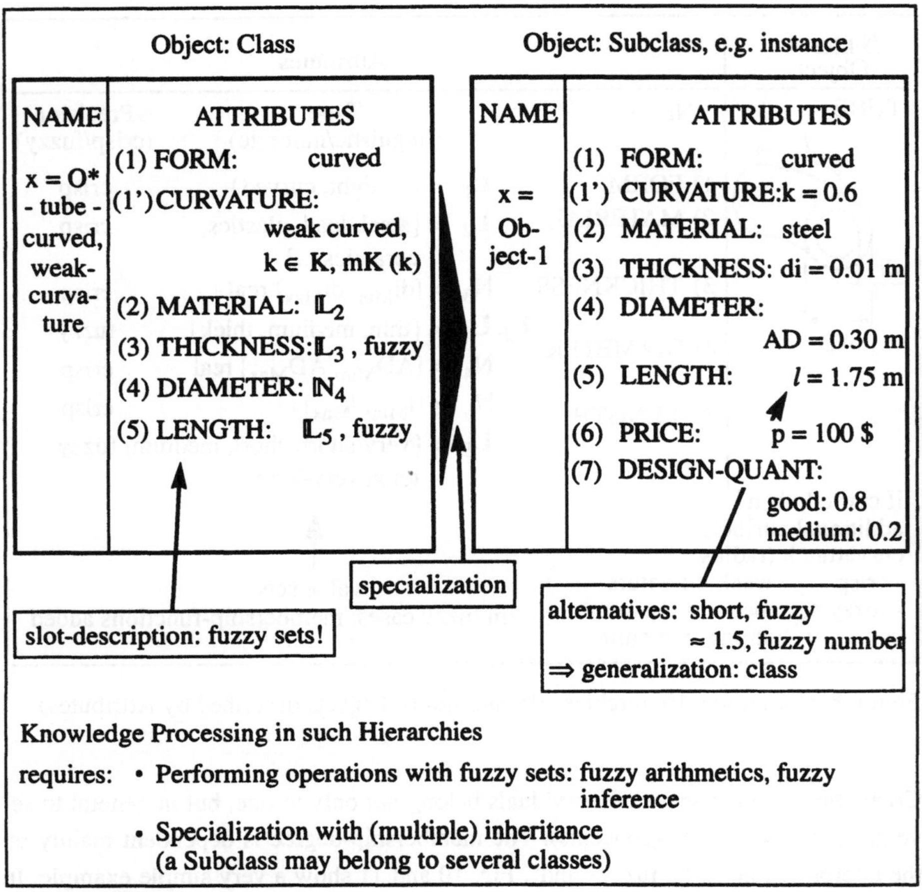

Figure 9: Classes - Subclasses

The compositional hierarchies (part-of, has-parts) (Fig. 6) can also be modelled using fuzzy methods, taking into account the problems that the component structure might be partly unknown in earlier design stages or that there are "optional" system components. Therefore, the components of an artifact can be defined as fuzzy sets where the membership function represents the "desirability" of a component to belong to the artifact or not. For more detailed considerations of this topic see [42].

Fig. 12 shows another illustrative example. We consider an object hierarchy of couches, where 3 imprecise attributes - Size, Color, Price - are considered. For the AND-Operator between those attributes not only the Minimum but also other operators are used [54]. Of course, the fuzzy sets for *small* - *medium* - *large*, *dark* - *light* and for

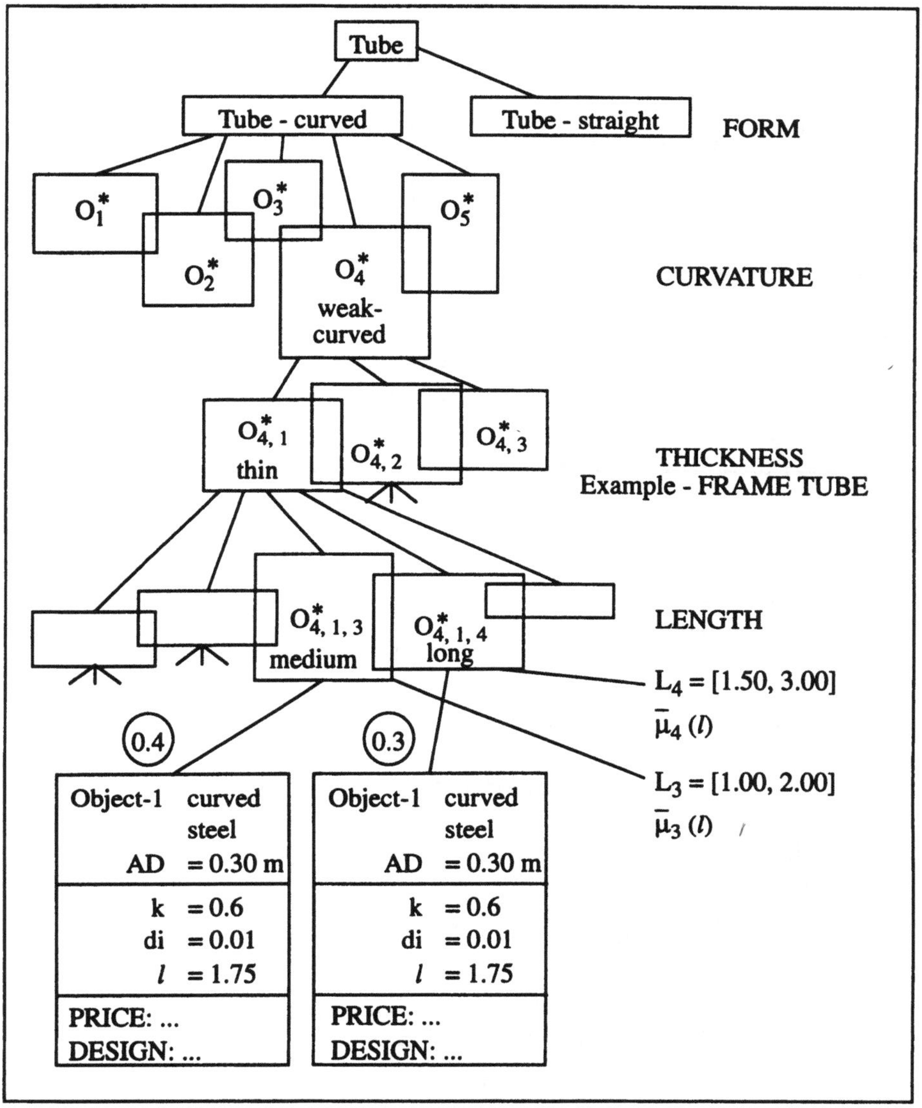

Figure 10: Example - FRAME TUBE

cheap - *normal-expensive* - *expensive* must be defined in the background of the picture. Now, if one has exactly defined objects like Couch-Conny (Size: 190 cm, Price 600 DM, Color: membership degrees related to *light* and *dark*) one might construct the hierarchy (thick edges) for the requirements Size: *small*, Color: *light*, Price: *cheap*.

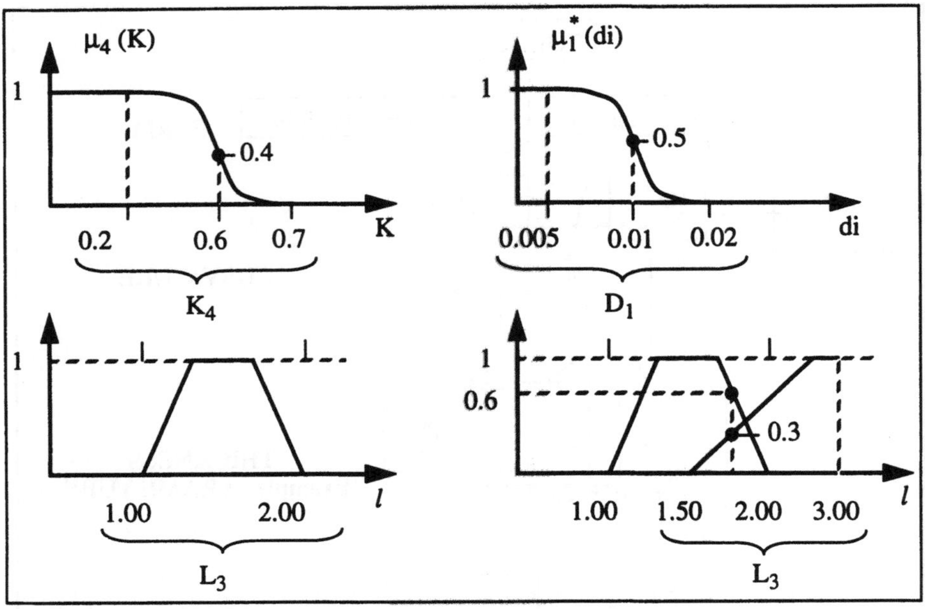

Figure 11: Specialization of a Class

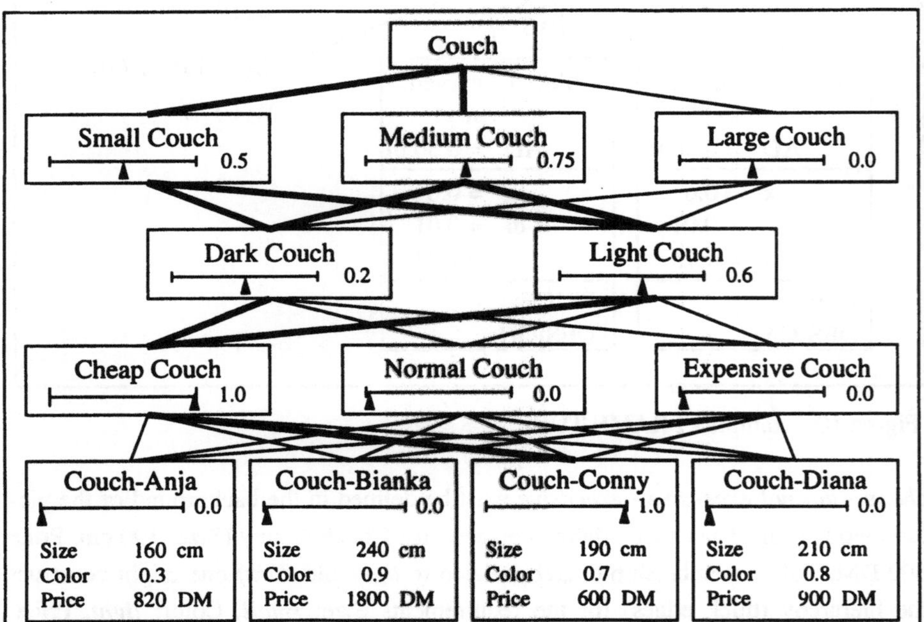

Figure 12: The Couch Example

2.2. The Requirement Model

2.2.1 The Crisp Case

One topic of central importance for a design or configuration model is an adequate representation of the user requirements. We consider three types of requirements:

- functional properties
- global constraints
- objectives

(1) Necessary Functional Requirements

With functional properties we model some requirements of functional abilities of the system, for example:

"The car should be fast and safe."
"The computer system should have graphic ability."

Now, we consider this linguistic expressions as crisp knowledge and we use such kind of requirements to specify the necessary components, object-instances the system must have. Later, we will alternatively model such kind of requirements taking into account the imprecision of the attributes used.

In the following, we give a syntactic definition of a necessary functional property F_j

$F_j = (f_j, (\{E_{j1}, .., E_{jk}\}, \{O_{j1}^*, .., O_{jk}^*\}, \text{key}(f_j)))j = 1, 2, .., \mu$
with

f_j : name of F_j

$E_{j1}, .., E_{jk}$ components, which must belong to S to realize F_j

$O_{j1}^*, .., O_{jk}^*$ corresponding to E_{jv} object-classes, e.g. object instances (defined explicitly or by constraints)

Then $F = (F_1, F_2, .., F_\mu)$ is called necessary functional requirement.

(2) Global Constraints

In the following, we use a vector x to describe the feasible objects for S or E_j as well and X to denote the set of feasible objects defined above by $\mathbb{P}$, $\mathbb{E}$ and GO (S).

Furthermore, there are resources of limited availability to the whole system S, e.g.:

C_1: Weight(S) $\leq G_{Max}$

C_2: Energy-consumption(S) $\leq En_{Max}$

C_3: Cost(S) $\leq C_{Max}$

These resources must be distributed to the system-elements under the "global-constraints", for instance:

C_1: Weight$(\{E_1, E_2, .., E_n\}, R(S)) \leq G_{Max}$

A configuration x of S is global consistent, if

- the functional requirement F is fulfilled
- x is local consistent
- x fulfills the global constraints C_j, j = 1, 2, .., M

$X^* \subseteq X$ denotes the set of global consistent configurations. A global constraint C_j is syntactically defined by

$C_j = \{c_j, (E_{j1}, .., E_{jk_j}), g_j\}$, where C_j is the name of the constraint

$E_{j1}, .., E_{jk_j}$ are the components of E (S) which need the resource modelled by C_j and

$g_j : X \rightarrow \mathbb{R}^1$ with $g_j(x) \leq 0$.

A constraint C_j may be "hard" or "soft", with other words crisp or fuzzy (see 2.2.2)

$C_j \in \{C^h_\nu, C^s_\delta\}$.

Then, $C = (C^h_1, .., C^h_{\nu_0}, C^s_1, .., C^s_{\delta_0})$ with $\nu_0 + \delta_0 = M$ models the <u>global constraint requirement.</u>

(3) Objectives and Fuzzy Goals

Optimization means that there exists at least one preference relation on X or X*, respectively. We introduce here a special kind of fuzzy goal immediately to compare the crisp and the fuzzy goal. But, for a more systematic consideration of fuzzy goals see 2.2.2.

We consider the following case:

1) Objectives

- There exist A crisp goals (aims)

$Z_1, Z_2, .., Z_A$,

called objectives, with:

If $x \in X$ is local consistent, then $Z_i(x) \in \mathbb{R}^1$ is the uniquely defined evaluation of x with respect to the goal Z_i (for each i = 1, 2, .., A).

2) Fuzzy Goals

- There are B "imprecise aims"

$$W_1, W_2, .., W_B,$$

so called "fuzzy goals" with:
W_j is described by a linguistic variable LV_j

$$LV_j \in \{w_{j1}, w_{j2}, .., w_{j\rho_0}\} w_{j\rho} \text{ linguistic values (terms)}, \quad \rho = 1, 2, ..., \rho_0$$

Then, $M_{j\rho}$ = "$LV_j = w_{j\rho}$" is a fuzzy set on X, characterized by the membership-function

$$\mu_{M_{j\rho}}(x): x \in X, \text{ for each } p = 1, 2, ..., p_0$$

- An aggregated fuzzy goal W_j^* is defined by:

$$W_j^*: X \to \mathbb{R}^1 \qquad \text{with} W_j^*(x) = \phi_j(\mu_{M_{j1}}(x), .., \mu_{M_{j\rho_0}}(x))$$

$$\phi_j: 0, 1]^{\rho_0} \to \mathbb{R}^1$$

An objective/fuzzy goal requirement for S is a tupel

$$ZW = (Z_1, Z_2, .., Z_A; W_1^*, W_2^*, .., W_B^*) \text{ (closed world)}$$

and an objective/fuzzy goal variant for S is a tupel

$$ZV = (Z_{i_1}, Z_{i_2}, .., Z_{i_{n_1}}; W_{j_1}^*, W_{j_2}^*, .., W_{j_{n_2}}^*)$$

with $n_1 \leq A$, $n_2 \leq B$, i_α, j_β pairwise different,

$$i_\alpha \in \{1, 2, ..., A\}, j_\beta \in \{1, 2, ..., B\}.$$

We consider a simple example of this linguistic variable approach:

<u>Example:</u>

A design object (a table, a bicycle, a car) should be described by 3 fuzzy goals:

W_1: Style ∈ {bad, medium, good}
W_2: Weight ∈ {heavy, normal, light}
W_3: Price ∈ {expensive, medium, cheep}

There are 3 x 3 fuzzy sets over X:

M_{11} = set of objects with bad design, $\mu_{Style = bad}(x)$

...

M_{33} = set of objects which are cheap, $\mu_{Price = cheap}(x)$

Let $x^{(1)}$ - weight of an object denoted by x

$x^{(2)}$ - price of an object denoted by x

Then we can use the fuzzy sets in Fig. 13 to define the linguistic expressions

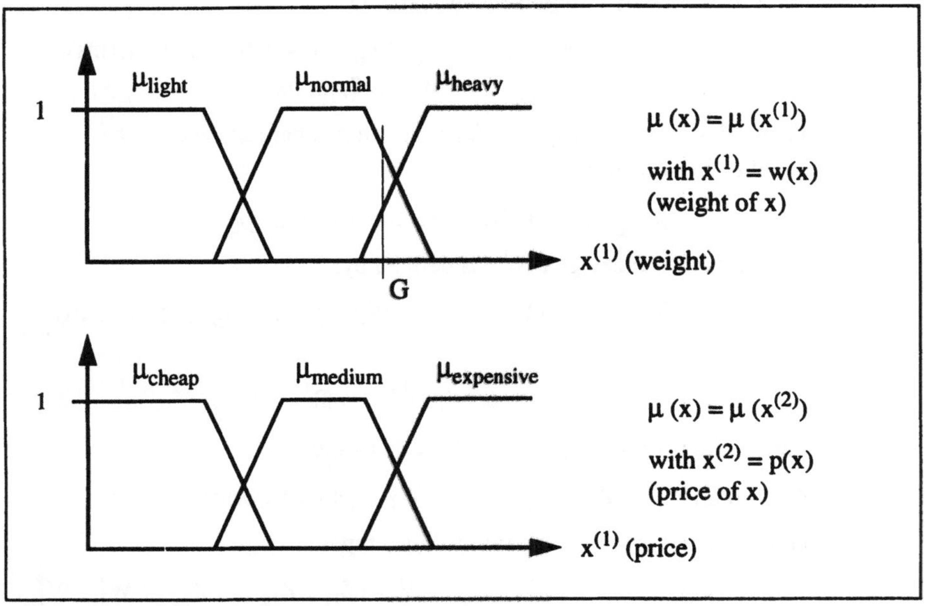

Figure 13: Fuzzy Sets for specific linguistic terms

Remarks:

All objects x with $x^{(1)}$ = W(x) = G are characterized by the same membership-function values. It is not obvious how to represent the membership-functions $\mu_{Style = ...}(x)$ depending on a real parameter $x^{(3)}$, $x^{(3)}$: [0, 1] which measures the degree of style quality. But, if for each x there is known a style-evaluation x^3 = st(x), and $x^{(3)}$ measures the style quality on [0, 1], $0 \leq x^{(3)} \leq 1$, ($x^{(3)} = 0$ absolutly poor, $x^{(3)} = 1$ perfect style) then the same situation as is in Fig. 13 occurs.

Now, the $W_i^*(x)$ might be defined by choosing functions Φ_1, Φ_2, Φ_3

$$W_1^*(x) = \Phi_1(\mu_{Style = bad}(x), \mu_{Style = medium}(x), \mu_{Style = good}(x))$$

$$W_2^*(x) = \Phi_2(\mu_{Weight = heavy}(x), \mu_{Weight = normal}(x), \mu_{Weight = light}(x))$$

$$W_3^*(x) = \Phi_3(\mu_{Price = expensive}(x), \mu_{Price = medium}(x), \mu_{Price = cheap}(x))$$

Later, we will define optimization problems. One problem would be to look for all non-dominated objects.

$$\begin{pmatrix} W_1^*(x) \\ W_2^*(x) \\ W_3^*(x) \end{pmatrix} \longrightarrow \text{Max* ("Pareto" maximum)}$$

We introduced fuzzy goals above using one special approach of modelling, the linguistic variable approach. However, the necessary functional requirements and also the global constraints are introduced, until now, only using the crisp point of view. Also, one might argue that the linguistic variable approach to the fuzzy goals defines a crisp substitute problem and does not deal with fuzziness.

Therefore, we now present a more systematic approach to the requirements in the case of imprecision.

2.2.2 Fuzzy Design Requirements

"Fuzzy-Design-Requirements" are modelling necessary or optional requirements or functional abilities for the goal-system, for example:

- The car should be fast and safe, show a modern style, but not be too expensive.
- The house should be designed not to be too energy-consuming.

First of all, these features allow the user's requirement to be formulated on the linguistic level. In 2.2.1 we used similar linguistic expressions, but we did not model the imprecision of such terms. Now, we mention that there is a linguistic approximation of the imprecise statements, e.g. "fast car", "energy efficient house", by fuzzy sets. Let x denote a car belonging to a set X of cars, ms = ms(x) (maximal speed) denotes a real valued parameter. Then, the fuzzy set, Fast-car = $\{x \in X \mid \mu(x) = \mu_{FC}(ms)\}$ is defined by a membership-function $\mu_{FC}(ms)$ (see Fig. 14). If for each car x the parameter ms = ms(x) is known, we call Fast-Car to be given explicitly. Otherwise $\mu_{FC}(ms)$ gives the possibility of each value ms, assuming the car is known to be fast.

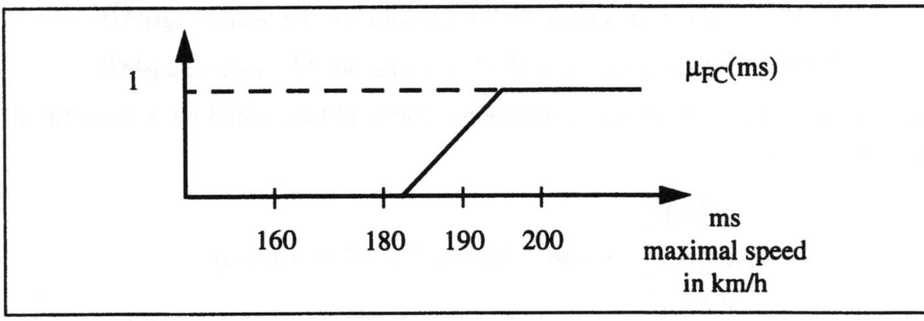

Figure 14: Fuzzy Set "Fast-Car"

This fuzzy set "Fast-Car" gives us a linguistic approximation using a quantification of "Fast" in the context of "Car". Of course, the example is very simple. But it is easily possible to make the example more realistic, e.g. to define Fast-Car by the fuzzy set Fast-Car = $\{x \in X \mid \mu(x) = \mu^*_{FC}(ms, t(0, 100))\}$ where t(0, 100) denotes the time the car needs to reach speed 100 km/h from 0.

It is more difficult but possible by introducing appropriate scales (possibly again multidimensional) to define "Safe-Car", "Modern-Style-Car", "Not-Expensive-Car" by fuzzy sets over X.

But, in addition, e.g. "Safe-Car" is connected with components this car should have and also specializations of these components which are more or less required (see 2.2.1). Therefore, Fuzzy Design Requirements are characterized by compositional hierarchies with imprecise attributes.

For instance, "Safe-Car", in addition to the fuzzy-set description (fuzzy set over X), means that the components "two independent brakes", "safety belts", "good tyres", for example, must belong, but one or two airbags only should belong to the car with a "desirability measure of e.g. 0.8". Also, ABS (anti-lock-brakes) could be a specialization-requirement for the brakes.

Fuzzy Constraints and Fuzzy Goals

The most important issue of fuzzy design requirements introduced above is to model the dependencies between an imprecise linguistic expression like e.g. "Safe-Car" and the component or specification structure which is required by this expression. However, such a fuzzy set might also be used to model the underlying requirement by a fuzzy-constraint or a fuzzy-goal. Let's consider the fuzzy set Fast-Car = $\{x \in X \mid \mu(x) = \mu_{FC}(ms)\}$ once more and assume that ms = ms(x) is known for each $x \in X$.

A crisp constraint could be formulated by

$$ms(x) \geq 190 \text{ km/h}$$

A fuzzy constraint is e.g. (according to the fuzzy set Fast-Car):

x is feasible with the degree $\mu_{FC}(ms)$, that means

$$\mu_{FC}(ms) = \begin{cases} 0, & \text{if} \quad ms(x) \leq 170 \\ \in [0, 1], & \text{if} \quad 170 \leq ms(x) \leq 190 \\ 1, & \text{if} \quad ms(x) \geq 190 \end{cases}$$

Using this kind of relaxation of the sharp boundary 190 km/h between feasibility and infeasibility we get a smooth transition.

There is another possibility to model the requirements as fuzzy goals, what means to have the idea of maximization of a real-valued function over X in mind, which measures the utility.

Again, let's consider the imprecise requirements Fast-Car, Safe-Car, Modern-Design-Car, Not-Expensive-Car represented by fuzzy sets over the basic set X of cars x. There are some different approaches to use such fuzzy sets to model fuzzy goals.

- The fuzzy goal is described by a fuzzy set. Then, we introduce a specific defuzzification by:

  - Find $x \in X$ such that $\mu_{FC_1}(x) \rightarrow \text{Max}$

    with $\mu_{FC_1}(x)$ as the membership-function of the fuzzy set
    FG_1 = Fast-Car $\wedge$ Safe-Car $\wedge$ Modern-Style-Car $\wedge$ Not-Expensive-Car

    e.g. $\mu_{FC_1} = \text{Min}(\mu_{F\text{-}C}(x), \mu_{S\text{-}C}(x), \mu_{MS\text{-}C}(x), \mu_{NE\text{-}C}(x))$

Still, there is another possibility to consider an aggregation tree (see Fig. 15).

We aggregate "fast" and "modern style" to "attractiveness". "Attractiveness" and "inexpensiveness" are aggregated to "rationality" and finally, "rationality" and "safety" will be aggregated to define the goal FG2. This aggregation uses a compensatory operator on the first two levels and the minimum operator on the final level [56].

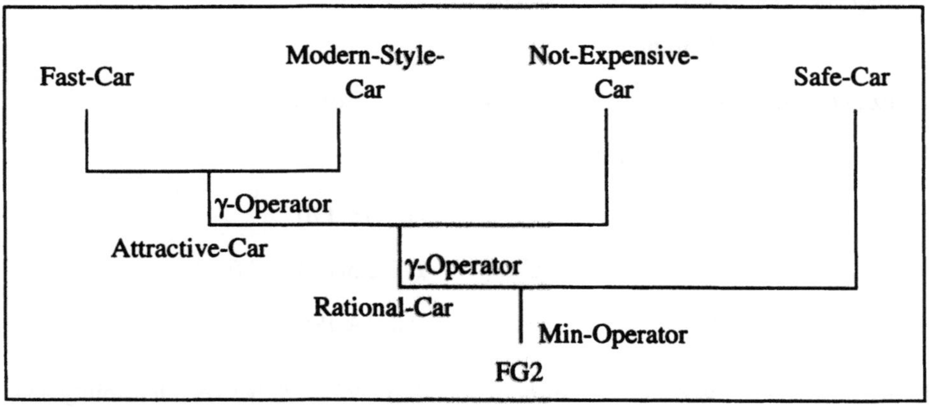

Figure 15: Example for an Aggregation

- Another possibility is the linguistic-variable approach to define fuzzy goals. We introduced this approach in 2.2.1 more mathematically. Here, we give another example. Let Maximum-Speed be a linguistic variable with the terms {slow, medium, high, very high}, then a fuzzy goal can be defined by "medium ∨ high". Then, again as a defuzzification, we are maximizing the membership degree of "medium ∨ high" over **X** using fuzzy sets "Max.-Speed = medium" and "Max.-Speed = high".

For more detailed considerations of the topics stressed, including the linguistic variable approach to fuzzy goals, see [41].

2.2.3 Specification of a Fuzzy Design Requirement Model

In Fig. 4 we introduced an Architecture of an IDSS for design and configuration tasks. Now, after introducing the basic concepts of the knowledge and the requirement model as well, we give an example of the "Requirement Modelling" and the "MCDM Model Generator" boxes of Fig. 4 in order to show how, in the case of fuzzy design requirements, this approach could be specified.

First, we provide the reader again with the terminology repeating the model parts from 2.1 and 2.2 we will use in the following:

- There is given a conceptual hierarchy of design objects, a compositional hierarchy and the set of conceptual constraints describing the domain knowledge of the design problem.
- We denote by

- FDR_j the fuzzy design requirement j defined by a name, a fuzzy set, a set of additional required components and a set of specialized objects
  e.g. target object "car"
  Fuzzy design requirement: Fast-Car defined by a fuzzy set over a crisp set X of cars using the maximal-speed as basic scale,
  required additional components: two airbags (desirability 0.8)
  required specialized components: ABS, special tires
- FC_k the fuzzy constraint k, where $k \in K$, K index set, if FC_k is a fuzzy set and μ_{FC_k} defines the "degree of feasibility" of an object represented by x.
- FG_l the fuzzy goal l, where $l \in L$, L index set, if FG_l is a fuzzy set and μ_{FC_l} describes a "degree of desirability" of an object x.

In the sense of a closed world, we assume the sets K and L to be known. That means all possible candidates for fuzzy constraints or fuzzy goals have to be predefined. Finally, a Modeling Support System helps to decide, which of the candidates will be used as fuzzy constraint or fuzzy goal respectively.

In the following an algorithm which realizes this conception will be described.

Configuration of the Model

Algorithm to configure the Fuzzy-Multi-Criteria Design Model

① Choose the target (goal) system, (e.g. car), from the conceptual hierarchy.

② Select the Fuzzy-Design Requirements FDR_j, in particular the fuzzy sets F_j from the set of all attributes of the target system (object) in the conceptual hierarchy

- Define J as index set of the FDR_j, $j \in J$.
- Determine the set of additional components and specialized objects related to FDR_j for all $j \in J$. (use the conceptual constraints, check inconsistencies, solve conflicts)

③ Select such linguistic expressions from the set of all attributes which are candidates for fuzzy constraints or fuzzy goals. (Use the predefined sets K and L. The F_j, $j \in J$ are automatically candidates for fuzzy constraints or goals as well.)

Define for an initial-model the fuzzy constraints and the fuzzy goals

FC_k, $\quad k \in K^* \subseteq K$

FG_l, $\quad l \in L^* \subseteq L$

④ Generate the "Fuzzy (Multi Criteria) Design-Optimization Model" by choosing one of the following approaches:

- Bellman-Zadeh-Model (if $K^* \neq \emptyset \wedge L^* \neq \emptyset$)
  ($F = (\text{AND } FC_k)$ AND $(\text{AND } FG_l)$, $k \in K^*$, $l \in L^*$)
- Defuzzification approaches
  - choose $x \in X'$ with max $\mu_F(x)$, (F, Bellman-Zadeh)
  - $FC^* = \text{AND } FC_k$, $k \in K^*$, fuzzy set over X'
    Define an α-cut of FC* and call $x \in X'$ α-feasible, if $\mu_{FC^*}(x) \geq \alpha$; $X = \{x \mid x \in X' \wedge \alpha\text{-feasible}\}$
    Define FL* by AND FG_l , $l \in L^*$ or by an aggregation tree (over several levels - AHP)
    Call x β-optimal, if x belongs to a β-cut of FL* over X (particular case $\beta = 1.0$).
- MADM (Fuzzy MADM) with
  X: finite set of explicitly given design alternatives
  (X' finite or X α-cut of FC* over X')
  FG_l , $l \in L^*$ set of imprecise attributes
  FG_l and the alternatives x_1 , ..., $x_n \in X$
  might be organized in a multilevel hierarchy (Fuzzy AHP), [52].

In step 4 of the algorithm several approaches to Fuzzy-MCDM are used. X' denotes the set of all partly feasible design objects x, that means $x \in X'$ fulfills the conceptual constraints (basic constraints) defined in the knowledge base. This algorithm is illustrated in Fig. 16. To identify the algorithm as a realization of the general architecture it becomes clear that the right hand part of Fig. 4 is still missing.

Therefore, a knowledge-based subsystem, which supports the "choosing one of the approaches" mentioned in ④ has to be added and presented (2.3, 2.4). Furthermore, the tool-box of solution methods for the different approaches (Bellman-Zadeh, Defuzzification approaches, Fuzzy-MADM approaches [5], [52]) developed as a part of the modular software system KONWERK will be explained in Sections 4 and 5.

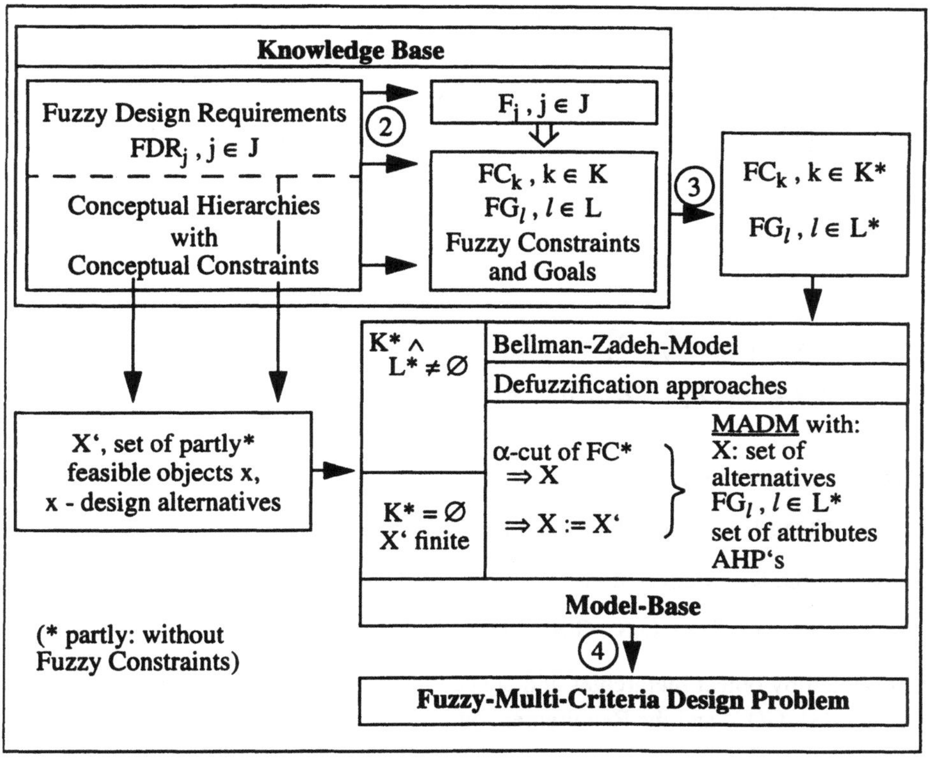

Figure 16: Algorithm to specify a Fuzzy Multi-Criteria Design Model

2.3 The Optimization and MCDM-Model

In 2.2.3., as an example we showed the specification of a Fuzzy Design Requirement Model (FDRM).

In order to summarize, we go back to the requirement model presented in 2.2.1., called the "crisp case" (but fuzzy goals of a very special type are included). Together with the "Knowledge Model" introduced in 2.1. we got the following class of configuration models.

$S \in \{S_1, S_2, ..., S_m\}$	Target system S belongs to a set of generic system-variants
$E(S) = \{E_1, E_2, ..., E_n\}$	Set of components (elements) belonging to S (see Fig. 6)

$\tilde{O}(S) = \{\tilde{O}_1(S), \tilde{O}_2(S), ..., \tilde{O}_n(S)\}$	Set of object sets related to S and to E_i, i = 1, 2, ..., n
$G(S) = (E(S), R(S))$, GO(S)	Structure and object structure graph of the target system S, where R(S) represents the binary relations between components of S
$F = (F_1, F_2, ..., F_\mu\}$	Necessary functional requirement of target system S
$C = (C_1^h, .., C_{\nu_0}^h, C_1^s, .., C_{\delta_0}^s)$	Global constraint requirement
$ZW = (Z_1, Z_2, .., Z_A; W_1^*, W_2^*, .., W_B^*)$	Objective I Fuzzy-Goal requirement

This configuration model class CONM

$$CONM = (S, E(S), \tilde{O}(S), G(S), GOS, F, C, ZW)$$

consists of the Knowledge-Model

$$KM = (S, E(S), \tilde{O}(S), G(S), GOS)$$

represented by object hierarchies and conceptual constraints (e.g. represented by the graphs G(S), GOS) and the Requirement-Model RM

$$RM = (F, C, ZW)$$

as well. It represents a closed world assumption and incorporates fuzziness in a more restricted way than the FDRM elaborated in 2.2.3.

This gives us the possibility to consider completely crisp design or configuration problems or to include fuzziness into the global requirement type or in the objectives.

2.3.1 An Objective/Fuzzy-Goal-Oriented Approach to the Solution of Configuration Problems

We assume that a configuration model of the model type CONM has been chosen. Then the task to find an "optimal system" S, which can be formulated as AND-OR-Problem Tree [4].

There are different possibilities - called perspectives - to genereate such a tree. Each perspective generates a special kind of algorithm [45], [46]. We consider an "optimization directed perspective" called "objective/fuzzy goal perspective".

The "objective/fuzzy goal perspective" means: we construct a decision tree with the following levels:

1)Decisions concerning objectives and fuzzy goals (level 1, 2, 3)

2)Decisions concerning discrete or continuous solution space (MADM or MODM) (level 4, 5)

3)Decisions related to methods to solve the problem considered in the subtree.

In Fig. 17 the construction of this tree is illustrated.

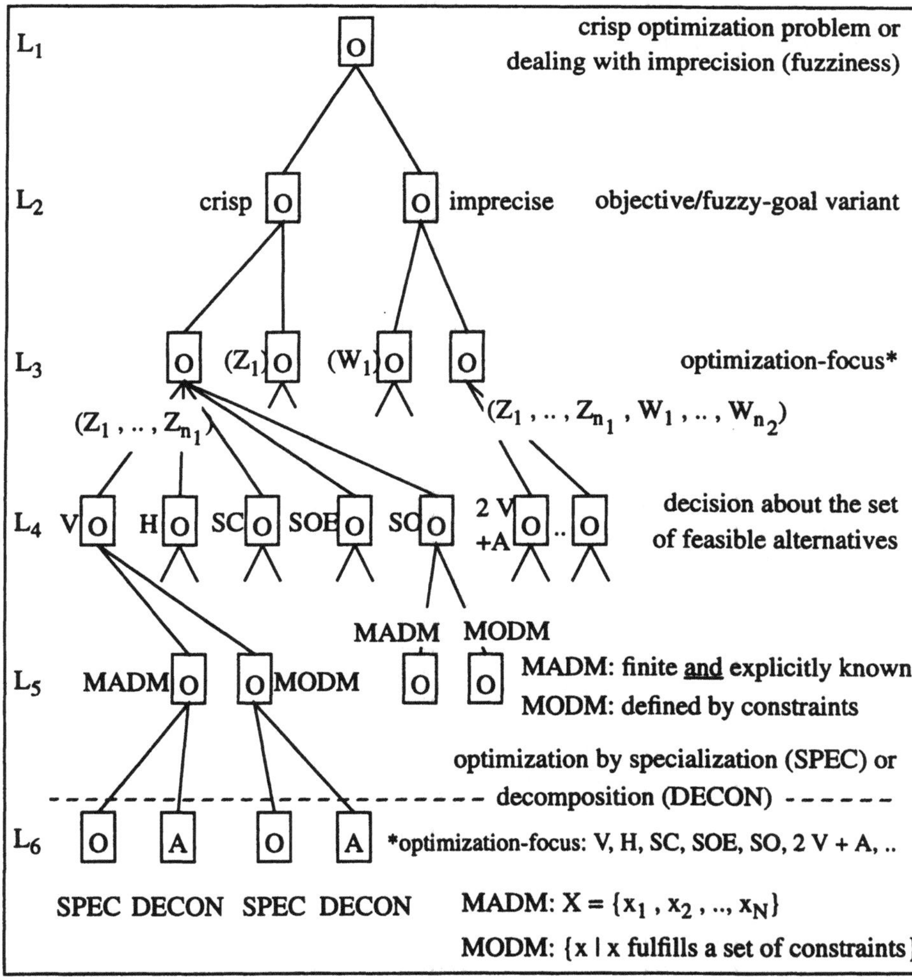

Figure 17: optimization-directed perspective

Explaining the optimization-focus V, H .. the following terms are used:

V -	vector optimization
H -	hierarchical optimization
SC -	substitute criterion
SOE -	single objective optimization and evaluation of the remaining criteria
SO -	single objective optimization only
2V+A	two vector optimization problems plus an overall aggregation of the criteria.

MADM (Multiple Attribute Decision Making) and MODM (Multiple Objective Decision Making) are explained in detail e.g. in [52].

Using this optimization-directed perspective we can define:

Optimization-based configuration means sequential top-down configuration according to the conceptual hierarchy, where each configuration step is defined applying the optimization-directed perspective.

Before considering this optimization-based configuration approach in detail, we describe some methods we need to give a compact formulation of the overall algorithm.

2.3.2 Basic Methods

In the following we define a method which generates an internal representation of the optimization problem under consideration.

(1) Name: OptbasPersp (Y)

generates: $(Z_1, .., Z_{n_1}; W_1, .., W_{n_2})$ (Y) — List of objectives/ fuzzy goals for Y

Focus (Y) — optimization-focus of Y

Set of feasible alternatives X (Y)
MADM: $X(Y) = \{x_1, .., x_N\}$ finite, explicitly known
MODM: $X(Y) = \{x \mid x \text{ fulfills a set of constraints}\}$

We assume:

- The configuration model (CONM) (or parts of it) acts as the input of a method denoted by OptbasPersp (Y) and of some other methods. That means, we assume that there is given the:

- conceptual hierarchy and conceptual constraints
- object-structure-graph (binary relations)
- global constraints (n-ary relations, if projected to components)
- functional requirements
- objectives and fuzzy goals

- Executing the method OptbasPersp (Y) we get:

  The set X (Y) of feasible alternatives for the concept Y of the conceptual hierarchy is obtained either by

  - a finite number of object instances of Y (specializations from the taxonomic hierarchy or caused by functional requirements)
  - a finite set of constraints associated with Y as "local" constraints or required from the functional requirements.

If Y = S, then the constraints associated with S are the global constraints. More formally we can describe this method by the name and the output generated.

The result of OptbasPersp (Y) is the optimization model Opt (Y) assigned to the concept Y.

If the optimization model Opt (Y) is generated by OptbasPersp (Y), one is able to look for methods to solve the problem formulated by Opt (Y). A first class of such methods is called SPEC (Y) which means optimization by specialization.

In the following we give a brief description of SPEC(Y):

(2)	Name:	SPEC (Y)
	input:	Opt (Y) Optimization-Model assigned to Y
	generates:	an instance or a set of instances (system-instances or component-instances) of the concept Y by optimization: - analysis of the properties of Opt (Y) - comparison between Opt (Y) and a set of optimization models represented in the knowledge-base (part: model-base) (property matching) - construction of an agenda of solution-methods (algorithms) - solving Opt (Y) using the top methods of the agenda - representing the solutions x (Y) of Opt (Y) (* x (Y): Name of an instance or parameter-vector *)

In Figures 18, 19 and 20 we give an illustrative example of the combination of OptbasPersp (Y) and SPEC (Y). In the first block <u>SubtreeVecOp</u> the method OptbasPersp (Y) generates Opt (Y). Then in the next level SPEC (Y) analyzes the properties of Opt (Y) (e.g. perfect solution, goal conflict, ..).

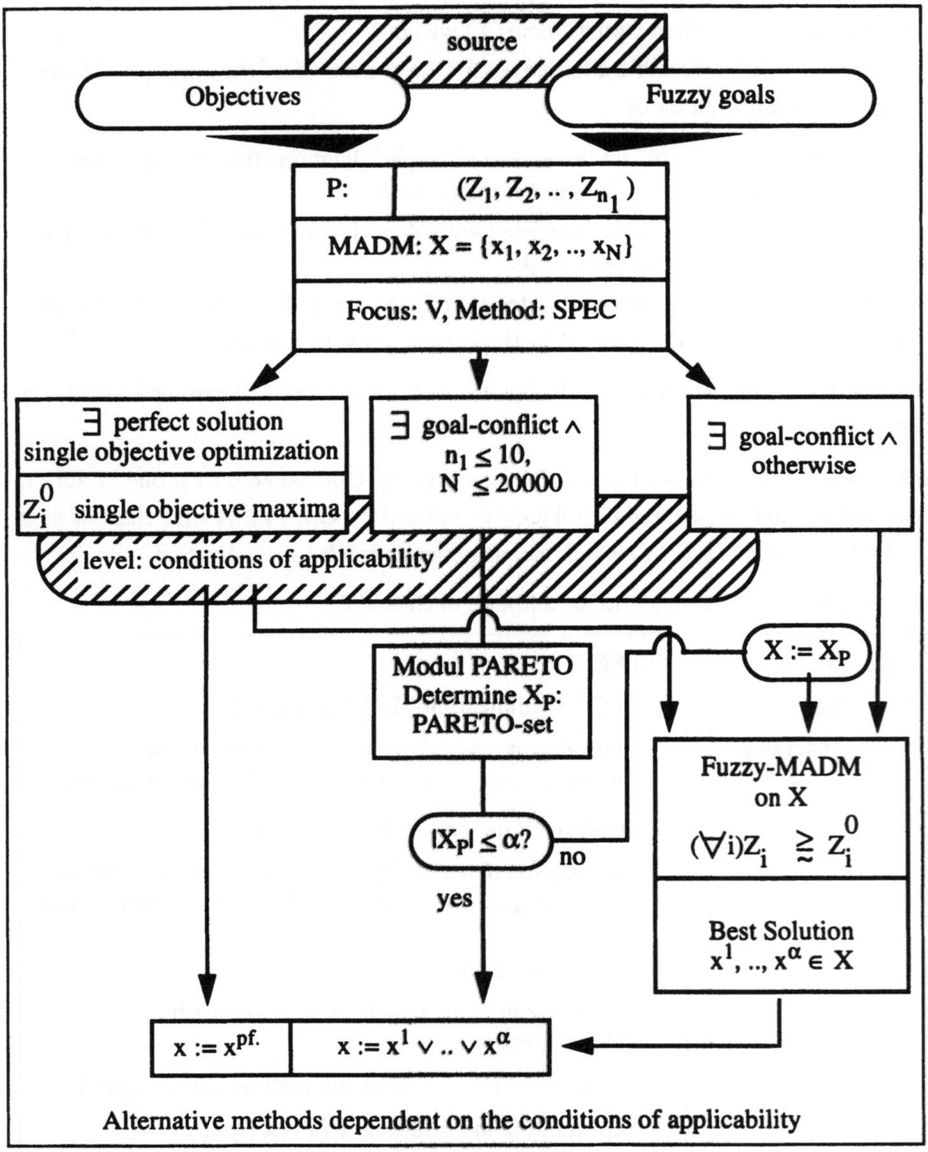

Figure 18: Subtree VecOp

Depending on these properties, different methods to solve the optimization problem by specialization (e.g. modul PARETO, modul FUZZY-MADM, ..) are available.

The special kind approach, illustrated in Fig. 18, is implemented and tested [8].

Optimization in the Case of Fuzziness

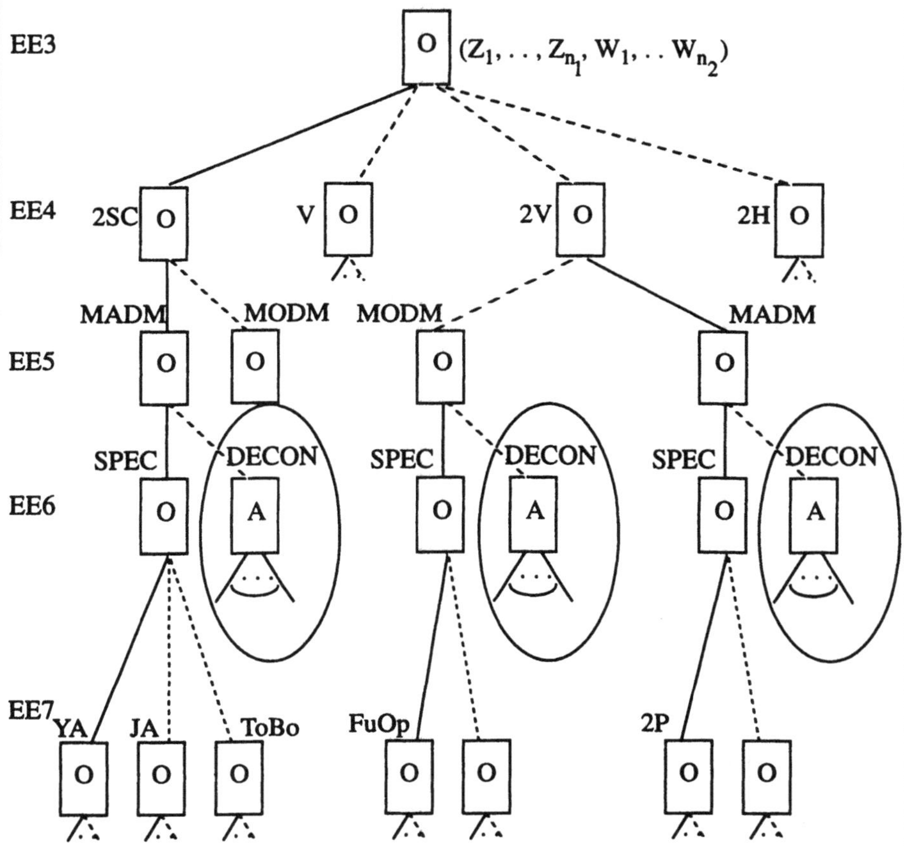

with:

YA	Method of Yager; aggregation of the W_j
JA	Method of Jain; aggregation of the W_j
ToBo	Method of Tong and Bonissone
FuOp	Fuzzy - optimization in the MODM-case
2P	2 Paretosets in the MADM-case

Figure 19: Subtree: Optimization in the case of Fuzziness, multi criteria, several realizations of SPEC

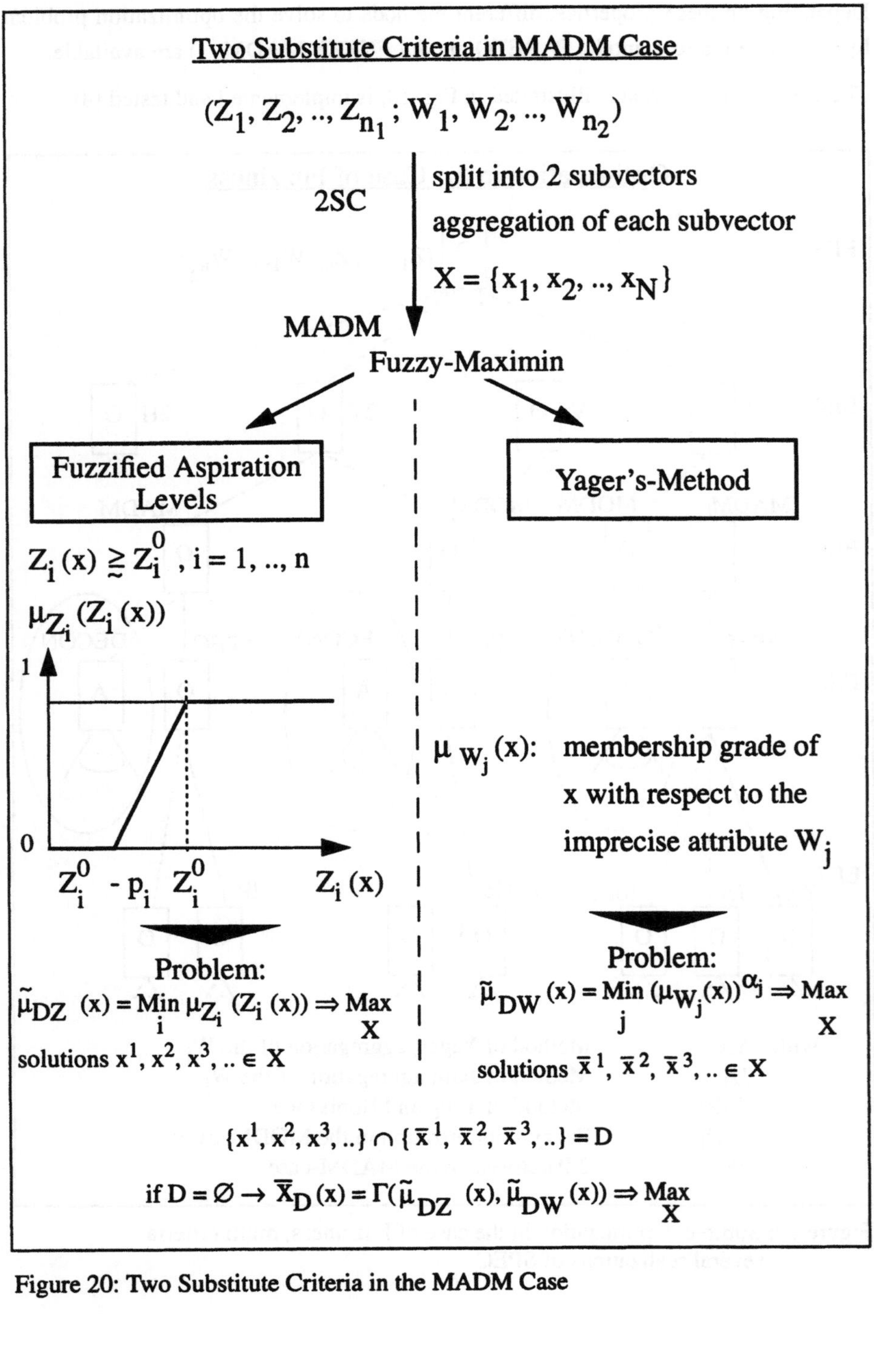

Figure 20: Two Substitute Criteria in the MADM Case

2.3.3 Decomposition-Methods

A second class of methods is devoted to the decomposition (DECON-methods). According to the compositional hierarchy it might be necessary to perform configuration by a decomposition-procedure. In the following we describe three methods of this type. The most simple approach is:

(3) Name:	DECON-1 (Y)
input:	- Opt (Y) generated by OptbasPersp (Y) - CONM, in particular the has-parts α-relation and functional requirements
generates:	- the components E of Y (let the components of Y be E_1 , .., E_n)
executes:	- OptbasPersp (E_i) for each E_i of Y (* projection-method for objectives/fuzzy-goals/global constraints according to the decomposition of Y *) - SPEC (E_i) for each E_i of Y - composition of a description X (Y) of the concept Y using an operator Γ applied to the component descriptions x_i (E_i). $X(Y) = \Gamma(x_1(E_1), .., x_n(E_n))$

There are two really interesting aspects involved. To be able to execute OptbasPersp (E_i) it is necessary to have a projection-method for objectives/fuzzy-goals/global constraints according to the decomposition of Y. Secondly, the solutions x_i (E_i) of the subproblems Opt (E_i) must be composed in order to find a feasible solution related to Y and we call this subtree a DECON-Node.

Projection and Composition of Objectives, Fuzzy-Goals, Global Constraints
We consider the subtree of the optimization-based perspective

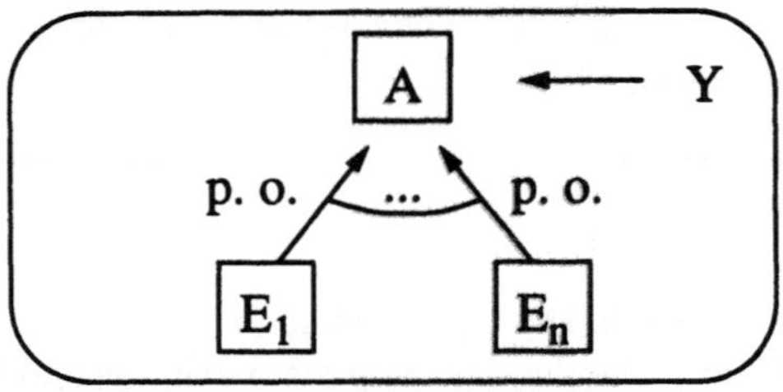

The method OptbasPersp (Y) generates e.g.:

$(Z_1, ..., Z_{n_1}; W_1, ..., W_{n_2})(Y)$

X (Y) represented by $G_1(x), ..., G_{n_3}(x)$ "global constraints for the subtree above (local for Y)"

$x = x(Y)$ description of an instance of Y

Then, projection is defined as follows:

Projection (e.g. for objectives and fuzzy-goals)

- Each component E_i has an objective/fuzzy-goal perspective

  $(Z^i_1, ..., Z^i_{n_{1i}}; W^i_1, ..., W^i_{n_{2i}})(E_i)$ with:

  - Z^i_k, W^i_r are projections of Z_k, W_r to E_i, $k \leq n_1$, $r \leq n_2$
  - Some of the Z^i_k, W^i_r might be empty (= NIL)
  - There may exist some other component-specific objectives/fuzzy-goals:
    $n_{1i} \geq n_1$, $n_{2i} \geq n_2$

The problems Opt (E_i) are solved by SPEC (E_i). After doing this, it is necessary to compose the solutions $x_i(E_i)$ of SPEC (E_i).

Composition

If the E_i's are instantiated by applying SPEC, ($x_i(E_i)$ are known), the concept-instance of Y associated with the DECON-subtree can be computed using an operator Γ (Γ: composition of the descriptions)

$x(Y) = \Gamma(x_1(E_1), ..., x_n(E_n))$

and

- The objectives/fuzzy-goals of Y are computable using composition operators Co_k, Co_r^*

  $$Co_k\ Z_k = \Phi_k\ (Z_k^1, ..., Z_k^n) \qquad k = 1\ (1)\ n_1$$

  $$Co_r^*\ W_r = \Psi_r\ (W_r^1, ..., W_r^n) \qquad r = 1\ (1)\ n_2$$

  (The composition operators might be represented by a set of fuzzy-if-then rules)
- There may exist objectives/fuzzy-goals, which are not computable by a composition operator from components (all Z_k^i, W_r^i equal to NIL).
- Projection of the constraints G_l (x) might be realized by G_l (Γ (x_1 (E_1), ..., x_n (E_n))

Such a projection-composition approach is the kernel of any DECON-method. However DECON-1 (Y) solves n isolated (separated from each other) problems, that means:
The object-structure graph and also the global constraints for Y are not taken into account solving the single problems Opt (E_i) by SPEC (E_i). Therefore, feasibility of the composed (by Γ) solution x (Y) is not guaranteed. A tree search algorithm with backtracking might be needed to generate at least one feasible solution x (Y).

Such a simple algorithm might be useful to apply, however, many inconsistant (for Y) solutions x (Y) may be generated. Therefore, the idea is to modify DECON-1 such that branches of the search tree, which will lead to infeasible solutions only, will be recognized as soon as possible and removed (cutted).

This modification called DECON-2 is briefly represented in Fig. 21. The most important property of this method in contrast to DECON-1 (Y) is that at each step the problem Opt (E_i) is solved and combined with an estimation of global feasibility.

That means an alternative is chosen to be a solution of Opt (E_i) only, if the partial configuration constructed up to this step has a chance (by estimation) to be completed to a global consistent configuration for Y. This kind of DECON-2 (Y) method is described in [39], [46].

A further example (class-of algorithms) of a DECON-2 method is published by Funke [9]. Funke uses the generation of the PARETO-set by method SPEC (E_i). He constructs then an agenda of non-dominated alternatives which have in addition a chance

Sequential (Configuration) Methods		Model-based parallel Configuration
<u>DECON-1 (Y)</u> SPEC (E_i) for all i (sequentially or in parallel), n-<u>isolated</u> problems that means: The constraints; - object-structure-graph, global (projected) constraints of Y - are <u>not considered</u> applying the SPEC (E_i)-method → Feasibility of the composed solution x (Y) is not guaranteed, backtracking may be helpful <u>OS-optimization-Skeleton</u> may be designed as • Depth-first search with backtracking (bounding of partial solutions identified as infeasible) • Branch and Bound Strategy • Best-first search using A* • Tabu Search Strategy	<u>DECON-2 (Y)</u> • generates the components E of Y and chooses (using heuristic knowledge) a specification (instantiation) sequence {E_i} i = 1, 2, .., n • Initialization of the object-structure graph and the constraints associated with Y • For i = 1, 2, .., n do: OptbasPerp (E_i) (* X (E_i) local defined set of alternatives *) SPEC (E_i) with The chosen alternatives must be consistent with the object-structure graph and the partially computed global constraints • Composition ...(analog DECON-1) if there is an aggregated function of the objectiveslfuzzy goals	<u>MOCON (Y)</u> • generates the components E of Y and the vector x (E) of <u>variables</u> • OptbasPersp (Y) (* constraints depending on x (Y) *) • OptbasPersp (E) for all E of Y (* Local constraints depending on x (E) *) (* Computation <u>in parallel</u> *) • Overall-model MO (Y) $x(Y) = \Gamma(x_1(E_1),..,x_n(E_n)) = (\bar{x}_1, \bar{x}_2, .., \bar{x}_N) = \bar{x}$ vector of variables instantiating Y constraints, objectives, fuzzy-goals depending on $\bar{x}$ • Solution of Opt (Mo (Y)): (* Find $\bar{x}$! *) <u>Examples</u> • Depth-first search with: - SPEC (E_i) generates a set of n - on-dominated alternatives (PARETO) consistent with the object structure graph and the global constraints (Y). - Elements of the Pareto-set are ranked by an aggregate-criterion in order to get an agenda, backtracking permitted

Figure 21: Modifications of DECON-1

for global consistency and he permits backtracking if needed.

Another idea of DECON-2 (Y) is that the kernel of the method, called "optimization skeleton", is also configurable. Only the skeleton itself is fixed. But there are several possibilities to realize the skeleton. One of them is best-first-search with backtracking [15]. If, in addition there is an aggregated criterion for each component given, then a branch and bound strategy [20], an A* algorithm, or Tabu Search are available to fill the Skeleton.

The methods DECON-1 and -2 are sequential algorithms. They might be parallelized by considering different specification sequences for the components E_i . In particular when using DECON-2 it makes sense to consider different sequences, because some of them will quickly lead to a "suboptimal solution", but using some other sequences much more computational effort to find a solution will be needed. Alternatively, in Fig. 21, we formulate a so called "model-based parallel approach", where a model of Y will be projected to the components of Y (subtree). Doing this, we get an overall model where the variables denote the design decisions for the components. Solving this model, we get the vector of all the design variables at once - not the components sequentially.To illustrate the MOCON-Approach we consider a simple example below.

A system S may consist of the components E_1, E_2, E_3. Then, we have on the conceptual level

System S

$Y := S$

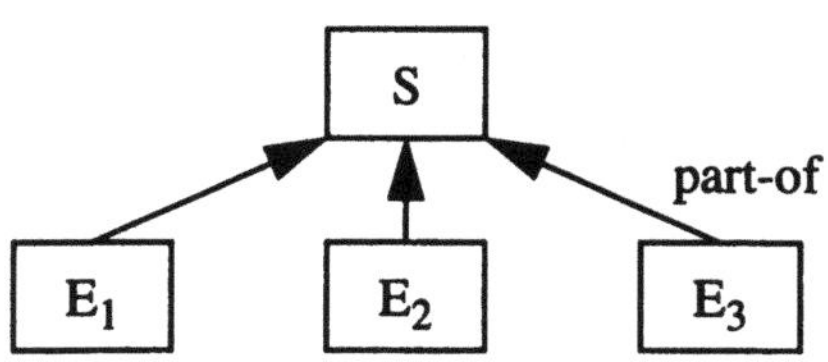

L_1

L_2

The description on level L_1 is:

$$x(S) = (\text{WEIGHT}(S), \text{COSTS}(S), \text{RELIABILITY}(S))$$

The constraints (global for the whole system S, but local from the perspective of level L_1) are:

$$x_1(S) = \text{WEIGHT}(S) \leq W_{Max}$$
$$x_2(S) = \text{COSTS}(S) \leq C_{Max}$$
$$x_3(S) = \text{RELIABILITY}(S) \geq Rel_{Min}$$

For each component we have a corresponding description:

$$x_i(S) = (x_1(E_i), x_2(E_i), x_3(E_i)),$$

with: $x_1(E_i) = \text{WEIGHT}(E_i)$

$x_2(E_i) = \text{COSTS}(E_i)$

$x_3(E_i) = \text{RELIABILITY}(E_i)$

for i = 1, 2, 3 and the local (level 2) constraints

$$x_1(E_i) \leq W_{\text{Max } i},\ x_2(E_i) \leq C_{\text{Max } i},\ x_3(E_i) \geq \text{Rel}_{\text{Min } i}\ .$$

Then, we have the following description-composition:

$$\begin{aligned} x(S) &= \Gamma(x(E_i), x(E_i), x(E_i)) \\ &= (x_1(S) = \sum_i x_1(E_i),\ x_2(S) = \sum_i x_2(E_i),\ x_3(S) = \operatorname{Min}_i\ x_3(E_i)) \end{aligned}$$

By renotation

$$\bar{x}_1 = x_1(E_1),\ \bar{x}_2 = x_1(E_2),\ \bar{x}_3 = x_1(E_3)$$

$$\bar{x}_4 = x_2(E_1),\ \bar{x}_5 = x_2(E_2),\ \bar{x}_6 = x_2(E_3)$$

$$\bar{x}_7 = x_3(E_1),\ \bar{x}_8 = x_3(E_2),\ \bar{x}_9 = x_3(E_3)$$

we get the following overall-model in terms of the design-variables for the components:

Find $\bar{x} = (\bar{x}_1, \ldots, \bar{x}_9)$

with $0 \leq \bar{x}_1 \leq W_{\text{Max } 1}, 0 \leq \bar{x}_2 \leq W_{\text{Max } 2}, 0 \leq \bar{x}_3 \leq W_{\text{Max } 3}$

$0 \leq \bar{x}_4 \leq C_{\text{Max } 1}, 0 \leq \bar{x}_5 \leq C_{\text{Max } 2}, 0 \leq \bar{x}_6 \leq C_{\text{Max } 3}$

$\bar{x}_7 \geq \text{Rel}_{\text{Min } 1}, \bar{x}_8 \geq \text{Rel}_{\text{Min } 2}, \bar{x}_9 \geq \text{Rel}_{\text{Min } 3}$

$$\left.\begin{aligned} &\bar{x}_1 + \bar{x}_2 + \bar{x}_3 \leq W_{\text{Max}} \\ &\bar{x}_4 + \bar{x}_5 + \bar{x}_6 \leq C_{\text{Max}} \\ &\operatorname{Min}(\bar{x}_7, \bar{x}_8, \bar{x}_9) \geq \text{Rel}_{\text{Min}} \end{aligned}\right\} \quad \text{global (for S) constraints}$$

In addition there might be several more constraints defined, e.g. binary constraints defining the structure graph (conceptual level) or also some other constraints caused by restricted available resources.

Examples are:

- A structure graph defining binary relations on the conceptual level:

If $\bar{x}_1 = x_1 (E_1) \leq \bar{w}_1$, then $\bar{x}_2 = x_1 (E_2) \leq \bar{w}_2$

(If E_1 is light, then E_2 must also be light)

After specification of the objectives and fuzzy goals, it is possible to apply a method of the SPEC-type to solve this model.

2.3.4 A Top Down Approach to the Global Optimum

In the following we propose a control strategy which is oriented to the conceptual hierarchy and which performs a top-down approach. It combines conceptual design (and if possible some kind of optimization oriented case-based reasoning) with design by decomposition-procedures. This approach may approximate the global optimum performing global and local decisions, because of the top-down orientation.

Several parts of this approach, in particular the model-manipulation-modules are quite imprecise and not yet formulated in detail.

Control Strategy (Top Down)
(Basic-Algorithm of Sequential Configuration)

1. Execute OptbasPersp (S)

2. If X (S) is finite and given explicitly,

    (* there is a finite number of existing system-instances, e.g. stored in a case base - library-solutions *)

   then execute SPEC (S) (* generates L (Opt (S)) *),
   else Goto Step 4.

3. If L (Opt (S)) is acceptable,
   then end

    (* global Optimum found by specification on the system-level - library-solutions *),

   else Goto Step 4.

4. (* X (S) is given implicitly by a set of constraints or L (Opt (S)) is not acceptable *)

   Using or generating of a representation of X (S) by a set of constraints
   Execute SPEC (S)

    (* SPEC (S) determines a $\hat{x} \in X$ (S), which is a solution of Opt (S), but $\hat{x}$ doesn't belongs to the library-solutions: Innovative Design *)

   If $\hat{x}$ is not acceptable

    (* (Z_1 ($\hat{x}$), .., Z_{n_1} ($\hat{x}$); W_1 ($\hat{x}$), .., W_{n_2} ($\hat{x}$)) is not acceptable under the focus chosen, 'the optimal solution' is not good enough *)

then Goto Step 6.
else Goto Step 5.

5. Execute DECON-2 (S)
(* Alternatives: DECON-1 (S), MOCON (S) *)
(* Depending on the chosen strategies for the optimization Skeleton - OS, DECON-2 (S) generates either a sequence of feasible solutions $\bar{x}$ (S) or an optimal (according to an aggregated criterion AG (x)) solution $\bar{x}$ (S) or finds out that there is no feasible solution *)

If there is inconsistency,
then goto Step 6,
else while DECON-2 (S) generates feasible solutions
If there is inconsistency,
then Goto Step 6,
else while DECON-2 (S) generates feasible solutions
Compare $\bar{x}$ (S) and $\hat{x}$ according to the objectivelfuzzy-goal variant and (if there is one) AG (x). Accept or reject $\bar{x}$ (S)

Choose the $\bar{x}$ (S) with the highest acceptance.
(* $\bar{x}$ (S) is the optimal realization of $\hat{x}$ *)

If all $\bar{x}$ (S) are rejected,
then Goto 7.

6. Execute Model Manipulation-1
If $\hat{x}$ not acceptable,
then repeat execute SPEC (S) until $\hat{x}$ acceptable or there are no other solutions of SPEC (S)

- Relaxation of constraints
- Relaxation of functional requirements
- Modification of the (F, C, ZW)-variant

The manipulated model Opt* (S) is used as input of SPEC (S)
Execute SPEC (S)
If there is no $\hat{x}$ accepted,
then end
(* Explanation of solution path and of conflicts *),
else Goto 5 using $\hat{x}$.

7. Execute Model Manipulation-2
(* DECON-2 (S) results in inconsistency or there is no $\bar{x}$ (S)

accepted *)

Generation of an agenda of components which are probably responsible for the failure

(* Expert Knowledge *)

Denote the top element of the agenda by E_{i*}

Relaxation of the local constraints of E_{i*}

Repeat until acceptance-test positive or agenda empty

- Execute SPEC (E_{i*}) (* generates $\hat{x}$ (E_{i*}) *)
- Construction of a solution $\tilde{x}$ for S by composition using $\tilde{x}$ (E_{i*})
- Acceptance-test of $\tilde{x}$ (S)

Fig. 22 gives an overview of the algorithm above.

However, to make this graph as simple as possible we did not use different symbols or marks to make the control flow completly clear. Therefore it is useful to consider both, the figure and the detailed description of the algorithm above.

2.4 A Hybrid Approach Branch and Bound Algorithm combined with Fuzzy MADM

The optimization-based control strategies introduced in this chapter are described on a relatively general level using complex blocks of operations. Complexity considerations, until now, have not been included.

We now introduce a two-stage approach, which is focused on a practicably usable algorithm for a class of problems which might be described as follows:

- There is a finite number of components of the target system which has to be configured
- Each component has a finite number of explicitly known instantiations (objects). For each component exactly one such object has to be assigned. We call such a complete assignment a configuration or a solution.
- There are constraints which determine whether a chosen configuration is feasible or not. These domain or requirement type constraints are assumed to be crisp.
- There is a finite number of crisp goals (objectives) Z_j , $j = 1, ..., A$.

With other words, we start with a specialized, completely crisp domain and requirement model as introduced in 3.2.1. and 2.2.

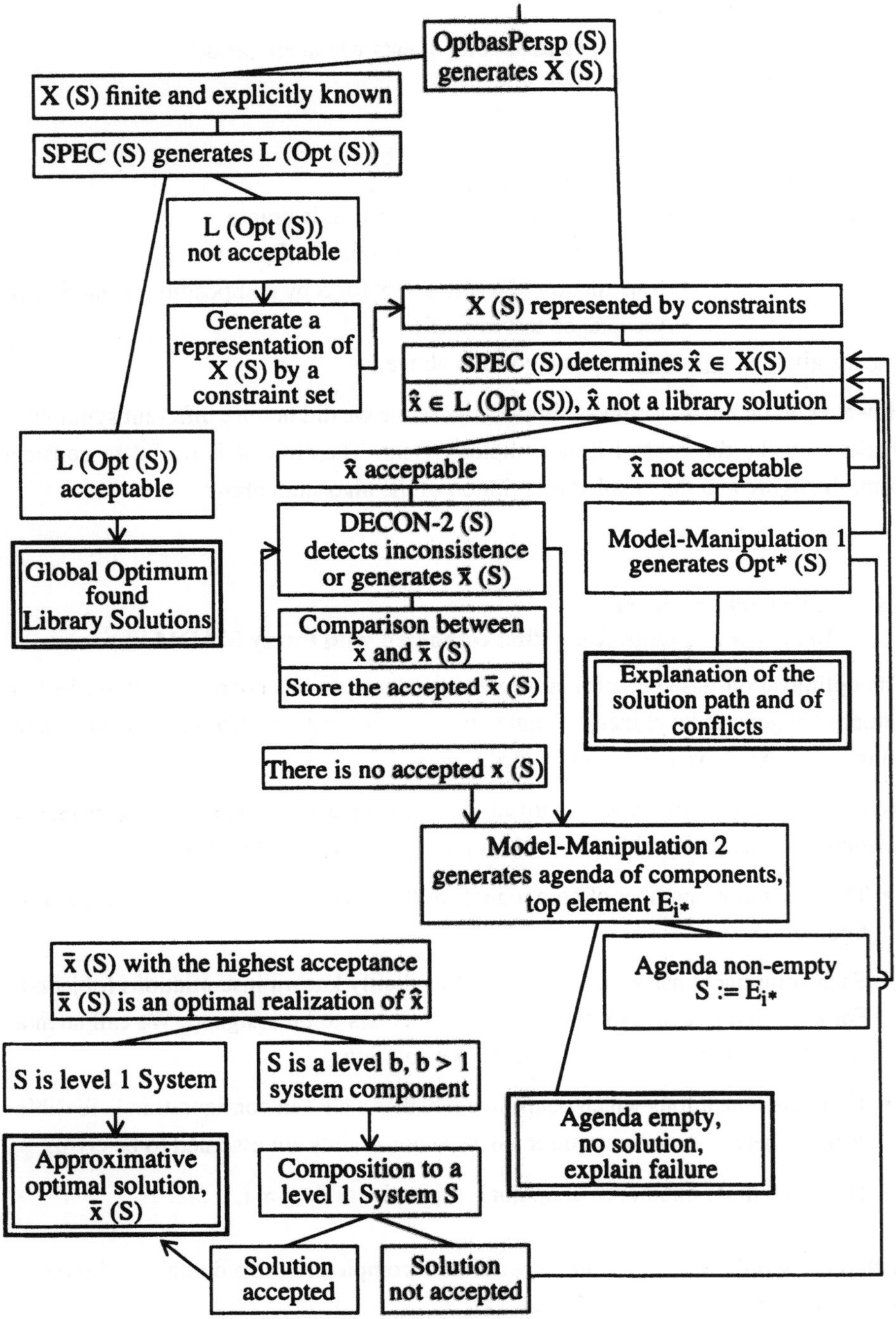

Figure 22: Basic Algorithm of Sequential Configuration

The Two-Stage Approach

First Stage: Generate, using a Branch and Bound Algorithm, a finite set of "good" configurations on the basis of the crisp domain and requirement model.

(This first stage might be modified to be applicable to imprecise domain and requirement models by introducing fuzzy Branch and Bound Algorithms.)

As a result of the "First Stage", we get a finite set of configurations. Now, we reformulate one part of the crisp goals into fuzzy goals and add, if needed, further fuzzy goals. Then, in the Second Stage, we apply a fuzzy-MADM method in order to rank and/or screen out configurations.

Second Stage: Apply a fuzzy-MADM method to the set of "good configurations" in order to find "optimal" configurations.

Description of the "First Stage"

The class of problems briefly described above can be modelled quantitatively as follows:

- The target system S is described by

  $S = \{E_1, ..., E_n\}$

  where $E_i = i = 1, ..., n$, are the components of S on the conceptual (generic) level.

- The object sets $\tilde{O}_i$ (set of specializations of E_i, generic or individual objects) are described by

  $\tilde{O}_i = \{O_{i1}, ..., O_{ik}, ... O_{iN_i}\}$

  with N_i explicitly known specialized objects O_{ik} of E_i.

For notational simplicity we assume $N_i = N$ for all i. If this assumption is not true, we define $N = \max_i N_i$ and define all objects O_{ik} for $k = N_{i+1}, ..., N$ to be infeasible (dummy objects).

As a result, we get an object matrix $(O)_{n,N}$:

specialization number / component		1	2	...	k	...	N
E_1	1	O_{11}	O_{12}	...	O_{1k}	...	O_{1N}
...	...	...	...	...	...	...	...
E_n	n	O_{n1}	O_{n2}	...	O_{nk}	...	O_{nN}

To model the instantiation process, we use decision variables $x_{ik} \in \{0, 1\}$, $i = 1, ..., n$; $k = 1, ..., N$, where

$$x_{ik} = \begin{cases} 1, & \text{if } O_{i,k} \text{ is chosen to realize } E_i \\ 0, & \text{otherwise} \end{cases}$$

That means, we have, in addition to $(O)_{n,N}$ a decision matrix $(X)_{n,N}$, where for each row the condition:

$$\sum_{k=1}^{N} x_{ik} = 1 \qquad (*)$$

holds, and each component has to be realized by exactly one of the objects O_{ik}, $k = 1, ..., N$.

If there are "optional" components E_i of S, then we get $\sum_{k=1}^{N} x_{ik} \leq 1$ instead of (*).

Using decision variables, the constraints defining the structure graph, (see [4]), might be represented as algebraic equations, see [20]. The decision variables can also be used to reformulate the crisp goals Z_j, $j = 1, ..., A$ and the requirement-type constraints (functional or recource capacity constraints). For a more detailed description see [21].

Now, to deal with the multi-criteria problem, we have first to explain more precisely, what kind of solution of the problem we would like to achieve.

There are (at least) the following 5 possible tasks:

(1) Generating a <u>set of feasible solutions</u> (configurations of the target system) and ignoring the goals completely.

(2) Generating a <u>set of feasible solutions</u> and simultaneously computing the values of the goals Z_j, $j = 1, ..., A$.

(3) Generating a <u>set of "good" solutions</u>.

(4) Generating a <u>set of "optimal" solutions</u>.

(5) Generating the <u>set of the m best solutions</u>.

<u>Task (1)</u>

The well established knowledge-based configuration approaches operate in this way by elaborating exactly one feasible solution [6]. More recent approaches seek a consisting of several alternative configurations.

Task (2)

In principle, this approach meets the requirements of task (1) and, after elaborating the set of feasible configurations, computes the values of the crisp goals for each solution belonging to the elaborated set.

Task (3)

First of all, of course, the term "good configuration" must be defined. S. Järvinen used well known approaches from the Multi-Criteria-Analysis field to define a solution to be good, if either:

1) an aspiration level is reached or

2) the difference to the utopia point is smaller than a given bound.

Aspiration level means that the user determines for each of the criteria a minimal (or maximal) value, which should be reached at least by a "good solution", e.g. the cost of a good configuration should be smaller than say $ 10,000 (cost is an objective one likes to minimize), but the reliability of the configuration has to be greater than 0.90 (objective to maximize). If for each crisp goal Z_j such a bound al_j (aspiration level) is defined, a "good solution" means a configuration which fulfills the respective inequality

$$Z_j \geq (\text{or} \leq)\ al_j$$

for each of the goals, $j = 1, ..., A$, (conjunctive method).

Another possibility is to define an aggregated goal

$$Z = F(Z_1, Z_2, ..., Z_A)$$

where $F: R^A \rightarrow R$ is an aggregation operator.

An aspiration level al for the aggregation Z is defined similarily to the al_j, but related to the aggregated evaluation of the solution (aggregation method).

An utopia point U is the vector of best (maximal or minimal) values $\hat{Z}_j$ of the goals Z_j. $\hat{Z}_j$ is computed as the optimal value of a single criterion optimization problem where Z_j is the objective function and the set of feasible solutions is defined by the whole set of constraints.

$$U = (\hat{Z}_1, ..., \hat{Z}_A)$$

If Z_j is a goal which should be maximized and the respective component of U is $\hat{Z}_j$ (maximal value of Z_j), then a "good" configuration should have a Z_j -value which has a less than α % difference to the utopia value $\hat{z}_j$:

$$z_j \geq \hat{z}_j - \alpha \frac{\hat{z}_j}{100}$$

Task (4)

In this case, the terminus "optimal solution" for a multi-criteria problem has to be defined more precisely. One might consider optimality in the sense of pareto-optimality (non-dominated set of solutions), or by considering an aggregated criterion or by the goal programming approach [52]. The aggregated criterion (often called utility function) approach and also the goal programming approach are directed to find a pareto-optimal solution which either maximizes an utility function or minimizes the difference to a desired (requested) solution (goal programming).

In the following, we will consider the case of one single criterion only or the case of a substitute criterion (utility function, goal programming) which also means to ultimately have one objective. This assumption makes the control-flow of the respective branch-and-bound algorithm more understandable. Generalizations are also possible [21].

Task (5)

In this case the idea is again to have one single criterion (possibly a substitute criterion). But now, the task is not determine the set of optimal solutions "only", but to generate the set of m best solutions. (If there are at least m optimal solutions, then task 5 means to determine a subset of the set of optimal solutions. But, if there is only one optimal solution and $m \geq 2$, then also the second best solution etc. must be derived).

Comparing the 5 tasks described above, task (1) seems to be the most simple and task (5) the most complex task. But, depending on the modelling decisions (the model builder has some freedom in choosing which are the constraints and which expressions are considered to be goals) the 5 tasks are not as different from each other than they seem to be.

Next we will focus on task (3), because we believe:

- the complexity of the resulting algorithms are still possible to handle (at least for classes of applied problems) and
- the resulting set of "good" solutions is the desired basis we need for fuzzy-MADM (second stage of the approach).

Both, the Branch-And-Bound (BAB) algorithm to determine a solution of a single objective optimization problem (task (4)) and the alternative BAB algorithm to elaborate a set of good solutions require techniques to deal with subconfigurations (not complete configurations, partial solutions). Therefore, before we present two BAB algorithms we give a brief introduction into approaches for dealing with partial configured systems (shortly called "subconfigurations").

A partial configured system consists of

- a subset of instantiated components ($x_{ik*} = 1$ for exactly one k*)
- a subset of non-instantiated components, so called free components

The respective index sets are $I_1, I_2 \subseteq \{1, 2, ..., n\}, I_1 \cup I_2 = \{1, 2, ..., n\}$.

Let's suppose the following structure of the crisp goals Z_j, j = 1, ..., A exists.

The evaluation of the component E_i by Z_j might be represented by

$$y_i^j = \sum_{k=1}^{N} c_{ik}^j \cdot x_{ik}$$

using the decision variables x_{ik}. Because of $\sum_{k=1}^{N} x_{ik} = 1, x_{ik} \in \{0, 1\}$ there is exactly one $k = k^*(i)$ such that $x_{ik*} = 1$ and $y_i^j = c_{ik*}^j$ if $i \in I_1$.

It follows: $Z_j = Z_j(y_1^j, ..., y_n^j) = Z_j(X_{n,N})$.

Let's assume that all goals Z_j should be maximized and that Z_j are monotonous separabel functions of $y_1^j, ..., y_n^j$. Then, it makes sense to use the following approach within a sequential configuration algorithm:

1) Initialization
   We initialize the evaluation of the component:

   E_i by $c_i^j := \max_{1 \le k \le N} c_{ik}^j$ for all j = 1, 2, ..., A

2) Updating
   For the instantiated components $i \in I_1$, the evaluation $c_{i,k^*(i)}^j$ are known.

Therefore these values are used instead of $\max_k c^j_{ik}$. For the free components $i \in I_2$, we use

$$c^j_i = \max_{k \in K(I_1)} c^j_{ik} \quad , i \in I_2$$

where $K(I_1)$ means that we consider only objects k which are still feasible taking into account that $I_1 \neq \emptyset$ is determined (structural constraints). If $I_1 = \emptyset$ updating equals initialization.

If the crisp goals do not have separability and monotonicity properties it will be difficult to find an approach with an appropriate "bounding" subprocedure.

Also, if there is some underlying structure of requirement type constraints, a similar initialization/updating approach can be introduced to check whether a subconfiguration might be completed to a feasible configuration or not [18].

1. Branch and Bound Optimization Strategy

We consider a task of the type described in task (4) with one single objective function (which might arise as a result of an aggregation procedure), and the optimization direction should be maximization. The data-input for the algorithm (Fig. 23) is the set of single evaluations c^j_{ik}, $i = 1, ..., n; k = 1, ..., N; j = 1, ..., A$.

The algorithm starts will the root-node of the configuration tree. It tries to branch any node as long as the node is not terminated (bounding). The instantiation of component E_i defines a level i of the tree.

The sequence of instantiation of the components is chosen to be fixed (expert knowledge) and enumerated by $i = 1, 2, ..., n$.

Comments:

Step 1: A node from the set of non-terminated (not bounded) nodes will be selected. There are several heuristics to select such a node.

Step 2: The bounding (terminating) of a node is realized by comparing the evaluation of the actual partial configuration with an existing lower bound.

Step 3: The branching strategy is a depth first search. A partial configuration will be pursued until a feasible configuration is elaborated or the partial configuration can be terminated. The complete configurations with the actual best evaluation are stored in L. If a better solution can be found, the lower bound has to be increased and the set L to be updated.

Step 0: Initialization

x_{ik} := 1 for all existing objects O_{ik}; x_{ik} := 0 otherwise.
Check the consistency of the root-node. If inconsistent, stop.
Generate a first lower bound $\underline{u}$ (initial solution or aspiration level)
Define: Set of non-terminated nodes M := {root}.
Define: Set of solutions L := { }. ({ } denotes the empty set also denoted by ∅)

Step 1: Choose the actual node

$M \neq \{\}$?
yes: select node $k^* \in M$
no: go to Step 4

Step 2: Bounding

Evaluate node k* by c_{k^*} : $c_{k^*} < \underline{u}$?
yes: Delete k* from M. Go to step 1
no: Check whether k* has free components or not?
- yes: i := index of the first free component of k*
  $M := M \cup \{k^*\}$. Go to step 3
- no: If $c_{k^*} > \underline{u}$ then $\underline{u} := c_{k^*}$
  $L := L \setminus \{k \mid c_k < \underline{u}\}$
  $L := L \cup \{k^*\}$. Go to step 1

Step 3: Branching

Does a successor l of k*, $l \in \{1, ..., N\} \setminus \{l \mid x_{il} = 0\}$ exist?
yes: Determine successor l, $x_{il} := 0$
Check consistency of l. If consistent then $k^* = l$. Go to step 2, otherwise repeat step 3
no: Delete k* from M. Go to step 1

Step 4: Optimal solution

Set of optimal solutions := L; Maximum of objective function := $\underline{u}$, Stop.

Figure 23: A particular Branch and Bound Strategy

Unfortunately the BAB algorithm has several disadvantages. Even if the algorithm finds an optimal solution very quickly, it might need a lot of additional effort to show that the solution found is an optimal one. In the worst case, the algorithm generates the whole configuration tree. Therefore, as an alternative, we consider now a modified

BAB strategy in order to obtain "good" solutions.

The main differences between a Branch And Bound strategy to find an optimal solution (introduced above) and a BAB strategy to generate a determined number of good solutions are the following:

- Lower bounds are the user-defined aspiration levels. There is no actualization of the lower bounds within the algorithm.
- The search procedure terminates if every node is terminated <u>or</u> if the desired number of good solutions is found.

In Fig. 23 the BAB-aspiration level strategy is defined.

For new problems, where the solution space is not explicitly known it might be difficult to define aspiration levels reasonably and also it might be difficult to determine in advance a number of required good solutions.

For example, if the aspiration levels are "too high", there might be no feasible solution which fulfills these requirements. If the number of required good solutions is bigger than the number of feasible solutions we would get a complete search (without success). The same would happen if the aspiration levels are low and the number of good solutions required is quite large. Therefore, an interactive BAB strategy was proposed by S. Järvinen [20].

We used much of effort to introduce a method for the first stage. The second stage of our proposed Two-Stage-Hybrid Approach is:

"Apply a fuzzy-MADM method to the set of good configurations".

It will be described in more detail in Section 4 together with the MADM and the fuzzy-MADM module of KONWERK.

From the first stage we know N* alternatives (configurations), which are explicitly given by the matrices

$$(X^i)_{n,N} \qquad i = 1, 2, \ldots, N^*$$

of decision variables. There is a guarantee of consistency. Therefore, we can easily assign names gc_i, $i = 1, 2, \ldots, N^*$ (gc_i - good configuration number i) to those alternatives. We also know the evaluation of Z_j of each of these alternatives: $Z_j(gc_i)$, and we know that the respective aspiration levels al_j are realized:

$$Z_j(gc_i) \geq (\leq)\ al_j \qquad j = 1, \ldots, A; \quad i = 1, 2, \ldots, N^*$$

Now we can use fuzzy sets to describe "desirable Z_j-values"

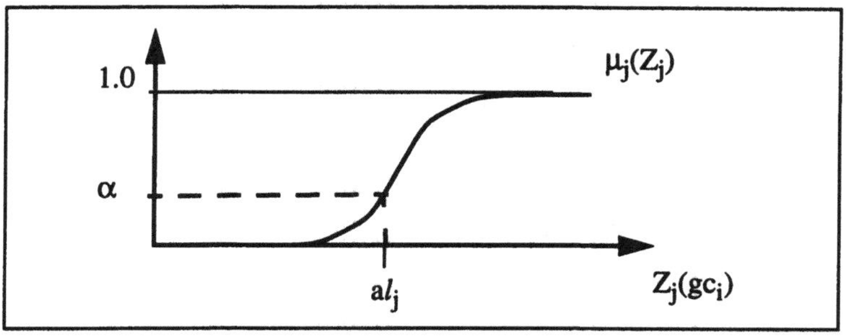

$\mu_j(Z_j)$ - defines the membership-degree to a fuzzy set of "desirable Z_j-values"
For each j, the aspiration level al_j should be defined such that $\mu_j(Z_j = al_j) = \alpha$.

This means that the crisp goals are fuzzified. Because for each gc_i, $i = 1, ..., N^*$, each crisp value $Z_j(gc_i)$ is known to be $\geq al_j$, each alternative has for each j a "desirability-degree" $\geq \alpha$. Now, we can also add other imprecise goals, lets say $W_1, ..., W_B$ (with $W_r = W_r(gc_i)$ a known real number, which has a degree of desirability).

Under those assumptions we are able to use fuzzy MADM methods to rank (or to screen out) the alternatives gc_i. These methods will be considered in Section 4. in connection with KONWERK. Also, in Section 5.2 we will give an example of the two stage approach. Finally, we remark, that using the Fuzzy Sets defined by $\mu_j(Z_j)$ we could also fuzzify our BAB-aspiration strategy by using not the conjunctive method with the Min-operator, but allowing compensations between the $\mu_j(Z_j)$, $j = 1, ..., A$ and a fuzzy number N^*.

3. Description of Applications

In the following we describe two of the five real-world applications we solved by applying hybrid type methodology introduced in Section 2 using a workbench of software tools (KONWERK) which will be described in Section 4. In this chapter we describe the domain and problem and in addition briefly outline the model and the solution algorithms only. KONWERK will be described in detail in Section 4.

3.1 Passenger Aircraft Cabin Layout

A cabin layout (shown in Fig. 24) defines the passenger cabin of an aircraft by number, arrangement, type and position of the cabin interior components. Cabin interior components are objects like passenger seats, lavatories, galleys or cabin attendant seats which may be installed in the passenger cabin. The entire set of cabin interior components in the passenger cabin defines its capacity, its comfort level, its possible service etc.

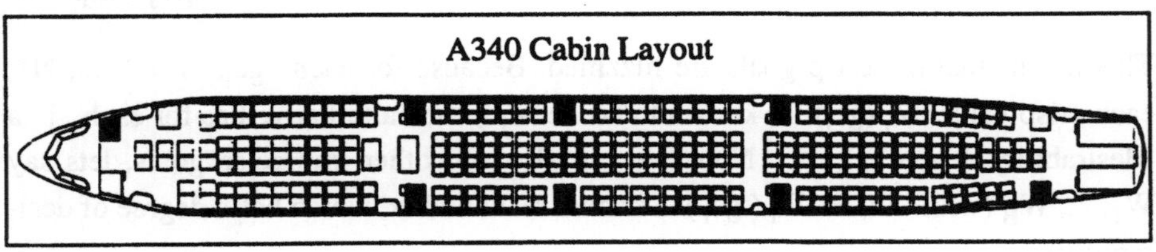

Figure 24: Cabin Layout that was configured with XKL

Configuring a cabin layout means [22], [23] to place a certain number of cabin interior components optimally in the restricted space of the passenger cabin. Optimality may be defined with respect to factors like the number of seats, the cost, the lead-time or a mix of these factors. In the process of configuring a cabin layout one has to consider the restrictions and the requests of the ordering airline. The restrictions may be either certification rules of the approving airworthiness authority or technical requirements of either the aircraft builder or the component supplier. The airline requests concern the division of the passenger cabin into passenger classes, the number of seats, the seat pitch etc.

The difficulty in configuring a cabin layout is to consider all restrictions and requirements at the same time. The following examples give an impression of the restrictions that appear in the cabin layout domain.

Figure 25 illustrates several certification rules: There has to be an assist-space to each side of a type-A-door (width at least 12 inch); the width of a cross aisle between two type-A-doors has to be at least 20 inch; a cross aisle between type-A-doors with two

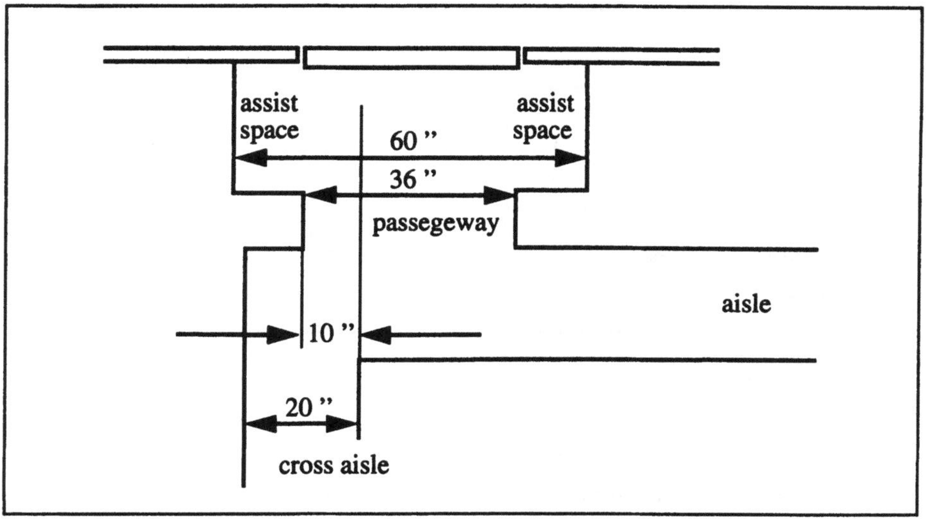

Figure 25: Restrictions for cross aisles between type-A- doors with two approach paths

approach paths has to overlap with the 36 inch of the passageway by 50%. Another example for a certification rule is the minimum ratio between cabin attendants and passengers.

There are also many technical restrictions, for example such as those for the mounting of a special galley. Another simple technical restriction defines the maximal number of lavatories in the passenger cabin.

In addition to the constraints (legal and technical) one has to consider the customers requirements. Figure 26 shows the seat pitch as an example for such a requirement.

One also has to strive for optimality with regard to criteria like number of seats, comfort level, cost, delivery time etc., which are different for each order. Configuring a cabin layout is getting harder. This is due to the constantly increasing size of aircrafts permitting more and more cabin interior components to be placed and a rising number of certification rules.

The aircraft builder's representative is no longer able to consider every restriction and thus cannot assure correctness and optimality of his layouts. An expert system might provide assistance by relieving the representative from routine tasks.It could make the necessary tests for correctness and thereby guarantee the consistency of the layout developer's decisions. Such a system should be able to carry out parts of the configuring

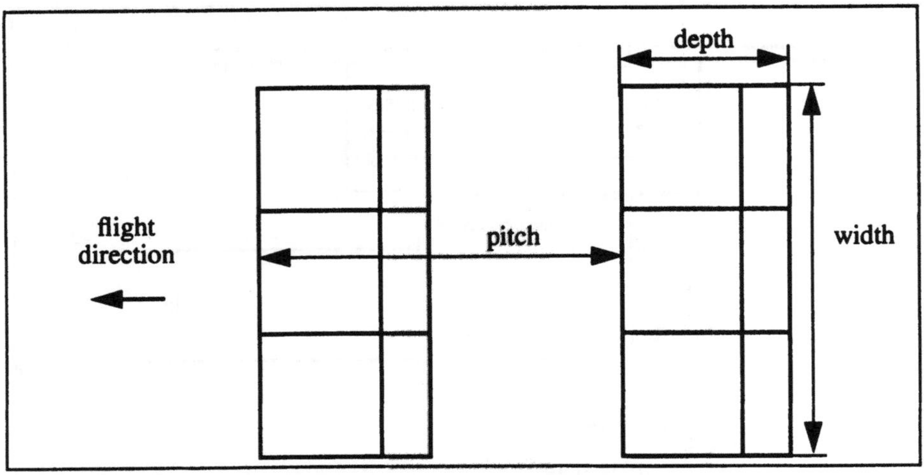

Figure 26: Customer's Requirement for a certain Seat Pitch

and optimizing processes on its own. The system XKL [22], an expert system for cabin layout, supports configuring a passenger cabin for an AIRBUS-A340. This system was build with PLAKON [6]. We used it as an outline of our considerations described in 5.1

3.2 Preliminary Design and Configuration of Future Space Launch Systems

In space flight a change of paradigm has taken place during the last years. While in former times national prestige and the 'cold war' were the main reasons for space activities, today economical aspects gain more importance. Furthermore, environmental compatibility and other requirements not only act like aims but like restricting arguments. The task of space launch systems (SLS) is to provide the transportation demand for orbital applications like positioning satellites in the orbit and planetary missions. SLS are the key technology for all space activities. Today a large number of different systems and a large amount of experience are available (Fig. 27). Future space launch system concepts depend strongly on the mission, i.e. the aims and requirements of the user. On principle, different technical solutions are feasible.

The research activities for new space launch systems are a time-consuming process taking years up to decades. The beginning of a project is especially important, because the main technical features will be determined in the first phase and cannot be changed in a latter phase. Technical and economic risks are determined in the early phase. The knowledge of experts helps to select technically feasible systems. The first

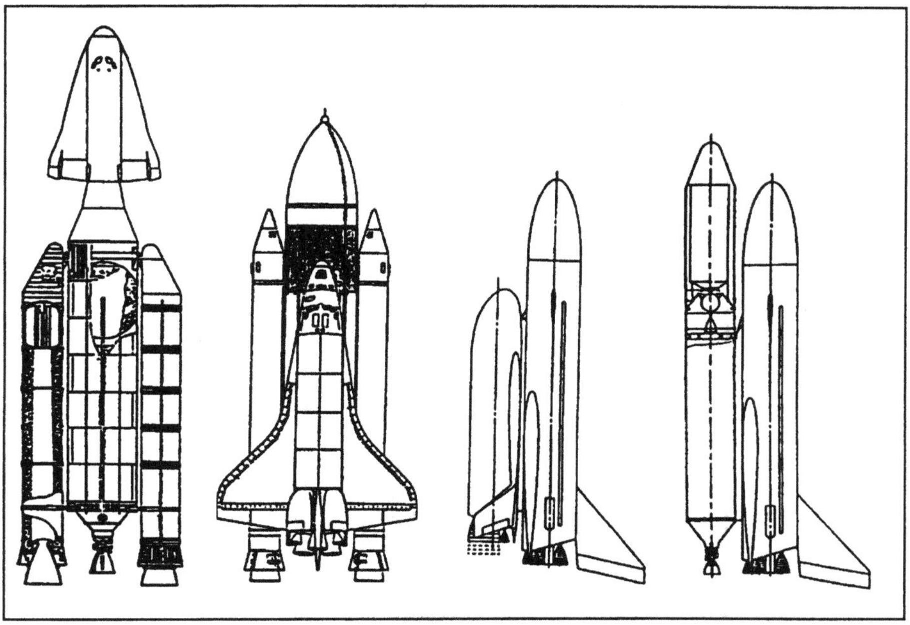

Figure 27: A selection of existing space launch systems

decisions give the main frame for the further research of the relevant systems.

In principle, the space transportation community is interested in a support system that assists the decision making process in the first approach towards a new space launch system. In the first approach a preselection of conceivable space transportation systems for a predefined payload with several requirements and one or more goals has to be made. The resulting configurations will be investigated more closely in a next step.

The first attempt to apply knowledge-based configuration in this technical domain was undertaken at the end of 1993 and was performed within a co-operation of the Institute of Operations Research of RWTH Aachen and the German Aerospace Research Establishment, Space System Analysis Division. The results have been presented in [13], [28], [29]. The implementation was done with the KONWERK workbench.

This Expert System assists in the decision process of configuration of future space launch systems. The highly complicated technical environment must be simplified, the same way an expert would. For a first approach it is sufficient to describe a space launch system with its components and several technical data. After a first selection, further optimizations have to be done, which is not in the focus of this work but could be an interesting additional question for an Expert System.

Space launch systems consist of a limited number of components, that will be aggregated modularly. This yields a good structure of the domain. The components are considered as concepts in a conceptual hierarchy. In this hierarchy both part-of relations and specializations can be described (see Fig. 28).

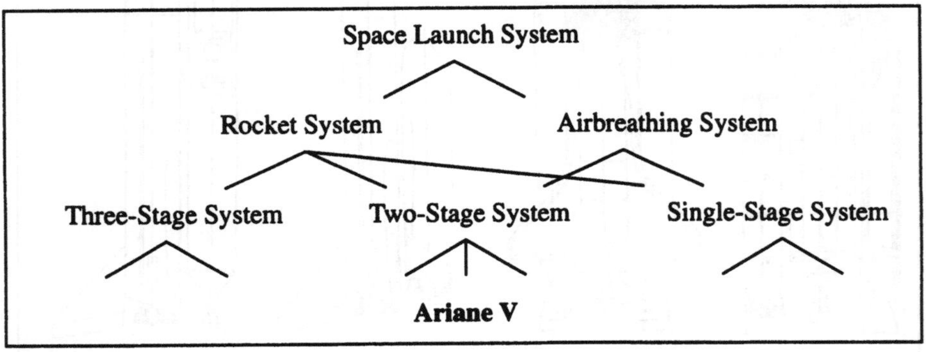

Figure 28: Specialization Hierarchy of Space Launch Systems

A space launch system is the most general concept and may be specialized according to the feature: first stage propulsion system. Further specialization concerning the number of stages is possible. In this tree-like structure the instances, which are the most specialized objects, are situated at the bottom, such as Ariane V, a two-stage, rocket propulsion system (see Fig. 29).

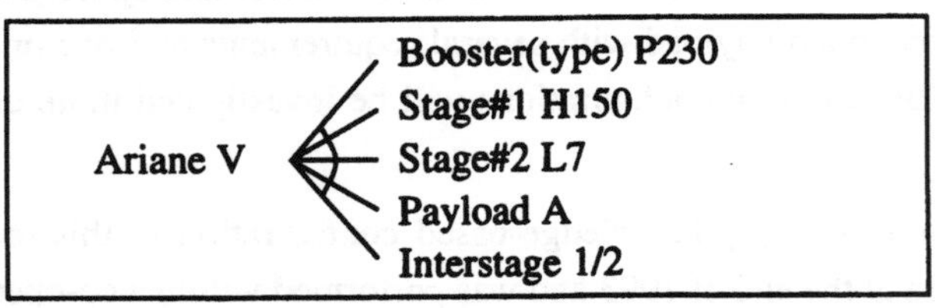

Figure 29: Part-of Hierarchy of Instance Ariane V

The concept Ariane V consists of parts that are instances of parts of general concepts, as well. They are specialized for the Ariane V system, and, therefore, determine this concept. Of course, the hierarchies in Fig. 28 and 29 are special parts of the overall hierarchy of objects which are related to the Space Launch (or Space Transportation) Systems domain.

The knowledge-base "SLS" (also called STS, see Fig. 30) is implemented in KONWERK and will be described later more precisely.

The design process starts with the payload mass to be transported, which corresponds to the payload capacity of the launch system. The aim is to configure a space launch system that is best adapted to the problem to bring a specified payload into a prescribed orbit. For simplicity reasons the payload capacity will be normed to the Low Earth Orbit (LEO).

In this example only rocket propulsion systems are considered. Airbreathing propulsion systems will be configurated similarly and, thus, will be neglected for this example.

The purpose of the first part of the design algorithm is to select those space systems, that are at least capable to transport the predefined payload into the orbit. This is the main part, since the number of components and the most important modules determining the space launch system are being fixed here. A set of numerical computations have to be performed to obtain the basic parameters, e.g. the take-off mass of the system. The geometric dimensions of the system are not known in the first step, these values will be estimated in a second step.

The algorithm generates a set of alternative SLS which are at least capable to solve the transportation task. Each alternative might be evaluated by several, in general imprecise criteria. Therefore, a Fuzzy-MADM can be applied to select or rank the alternatives.

For data input of the design algorithm crisp data is used presently. But the imprecise form, e.g., linguistic expressions, is a possible alternative. That makes the data handling easier for the user and helps him to formulate his requirements, which are imprecise, in a more natural way.

In the conceptual hierarchy, specializations that inherit characteristics of two superconcepts are likely, e.g., there exist plans for space launch systems with propulsion systems that are both, airbreathing and a rocket. In addition to this crisp values are not always wanted, and for the preliminary design phase fuzzy results are sufficient. An example is a fuzzy input of the payload capacity, that could be grouped into several weight classes, each class representing a fuzzy set.

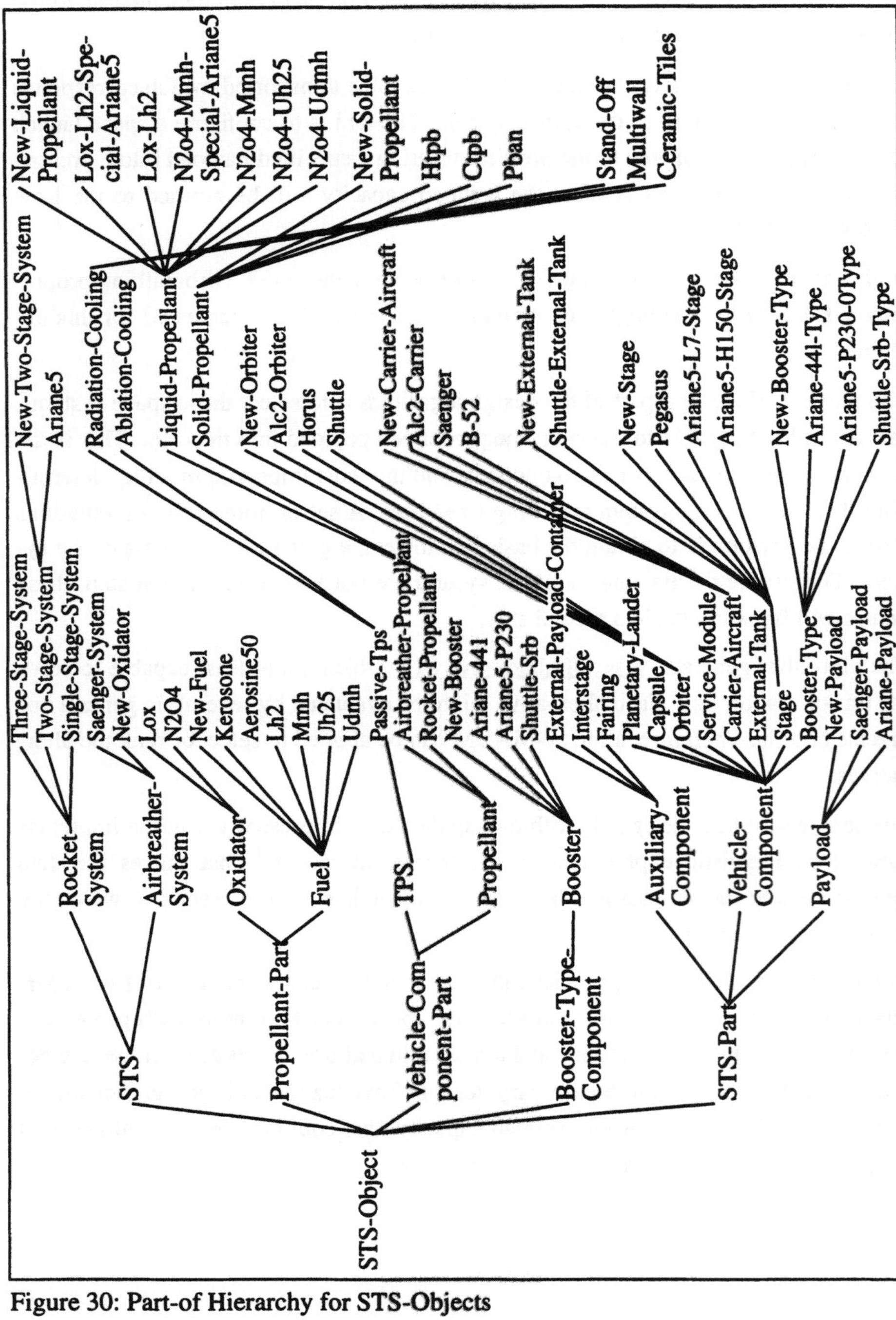

Figure 30: Part-of Hierarchy for STS-Objects

3.3 Logistic Layout

Another application is the so called "logistic layout problem". The task is to design a "transportation system" that consists of:

- several location points (e.g. distribution points, storages, receivers),
- transport relations (streets of a street network, air-transport relations, railway connections) between the location points
- transportation goods (homogeneous goods like oil or concrete, containers etc.)
- transportation vehicles (trucks, cars, aircraft, trains)

The locations might be producers of the transportation goods or distributors only. Some of the location points might be fixed (e.g. the producers, the final receivers and the distribution points, depots), some other might be variable, e.g. the so called hubs. (Hubs are storages where goods are collected, temporarily stored and then again transported often using a different type of vehicle.)

Now the task is to choose transportation vehicles and main relations, the variable locations (hubs) and the routes of the vehicles such that the whole transportation system fulfills its requirements. That means, it realizes the time constraints (also called time windows), capacity constraints (of hubs, storages, vehicles) and it maximizes (or minimizes) goals e.g.:

- minimization of investment and transportation costs
- maximization of the systems performance
- maximization of environmental compatibility

In Fig. 31 we show a simple example to illustrate such a system.

In this example the hubs H1 and H2 (airports) are chosen. H3 and H4 have been ignored (not used or decided not to build), and H5 was selected to change goods from fast large long distance trucks to smaller local trucks. The possibility of using railway transportation was not used in this example.

The question is now, how to get such a solution. In principle, it is possible to model the layout problem defined above by a mixed integer problem if all potential objects are assumed to be given. Using decision variables, one can model the decision about the kind of transportation used (air, ...) and about the hubs which should be used or not. The modelling of the possible resulting routes is well known in the Operations Research area. But, the resulting models are in general so big that it is almost impos-

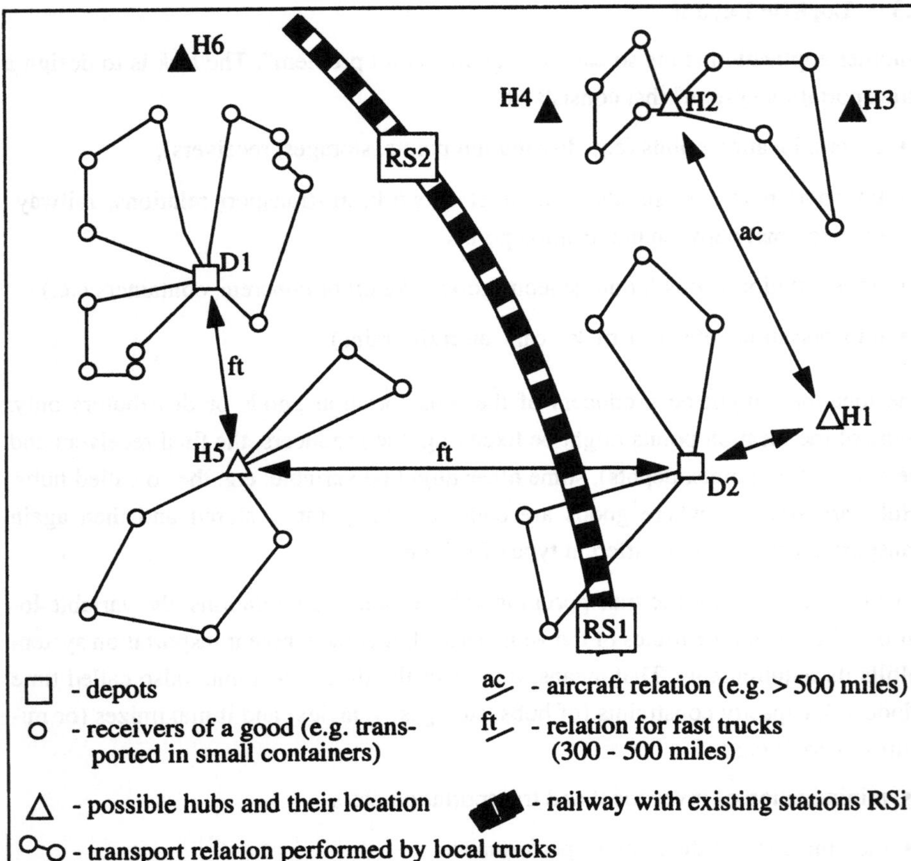

Figure 31: Example - Logistic Layout

sible to solve them. Usually several heuristics are used to find solutions of location-routing problems [26], [27]. To illustrate our approach, we will consider an example first and explain then the approach for this example.

<u>Example - Transportation of letters</u>

We consider a transportation system for collecting letters from mailboxes and transporting them using small cars and local trucks to a - lets call it - "letter centre of the town" (or the area). In some sense this system is inverse to the one described above because the goods (in this case the letters) are not distributed but collected. Of course, this is a small part of a much bigger logistics system ("LETTERS").

Summarizing, we have the following objects:

- mail boxes ○
- potential hubs △ (collecting letters), (△ means the potential hub is selected to be used, ▲ is not selected)
- the "letter centre" (depot) □
- transport relations with small cars○-○
- transport relations with trucks ◄─►

From Fig. 32 one might easily see that the trucks are used for routes where the letter centre is the depot and the used hubs are the "towns", if one compares it with a vehicle routing problem. Also at a second stage the hubs might be considered as depots and the (assigned to a hub) mail boxes are the "towns" for a vehicle routing problem where the vehicles are now small cars. For collecting letters there are tight time windows. It is not allowed to take the letters from the mailbox before the point of time which is announced at the mailbox to be the first possible time point to do this. On the other hand side the mail must be at the "letter centre" at determined time at latest to meet the system deadlines. But there are also other constraints. The mail boxes have different capacities and the capacities of the used vehicles are also different. Therefore the feasible routes for a vehicle are sometimes looking strange if one only considers the topology but not the time windows, the mailbox and the vehicle capacities.

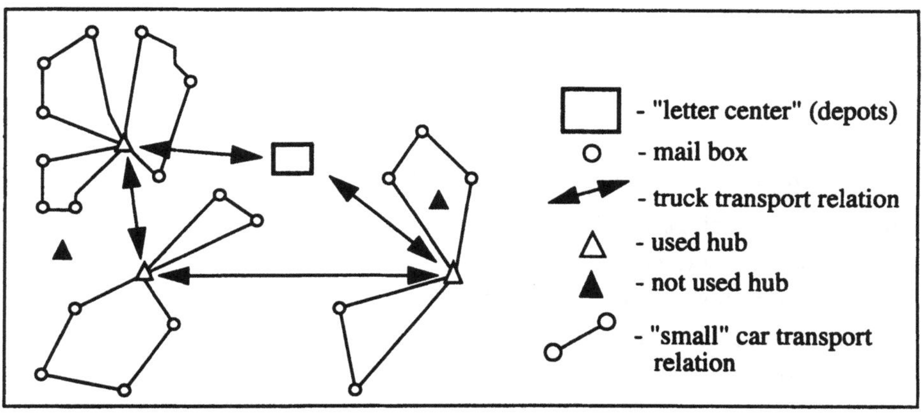

Figure 32: Transportation of letters - an example

Another problem is the fact that this transportation system has to be realized on an existing street network. Therefore, we use a GIS (Graphical Interface System) to be

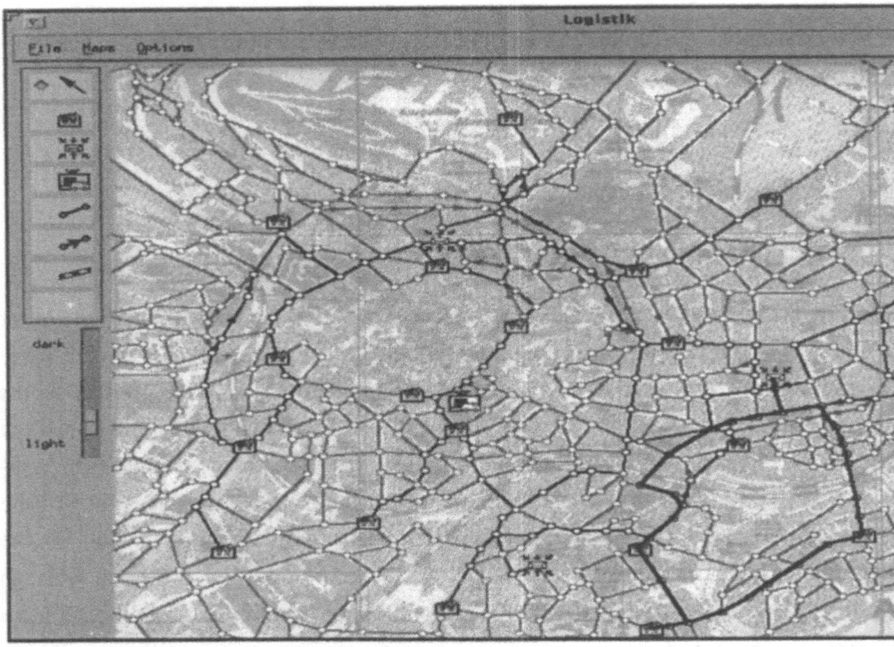

Figure 33: Graphical User Interface

able to determine the best transportation paths between locations taking car on "one-way-street" street types etc. In Fig. 32 and 33 there are shown both a formalized example using the symbols introduced above and one part of the graphical user interface.

3.4 Configuration of a PC (Personal Computer)

In the Artificial Intelligence (AI) area the knowledge-based configuration of computers might serve as a reference examples for configuration-type Expert Systems. We use the PC-configuration application to show a completely different approach compared to the approaches considered in Section 2 - the evolutionary algorithm approach. The implementation of the method described in the following is not coupled with the knowledge-based system KONWERK, but this will be done in the near future. The reasons we do experiment with evolutionary algorithms for design and configuration tasks are mainly the following:

- the configuration problems become very big because of complexity reasons,

- even if it is possible to model the multi-criteria optimization problem related to the configuration task appropriately, it is difficult to find optimal or good solutions by sequential type configuration approaches,
- for Fuzzy MADM methods we need to generate a set of alternatives which are feasible and which should be ranked afterwards by multiple criteria.

The evolutionary algorithms sequentially produce generations of populations and evaluate the individuals of these populations by fitness functions. Then, if the evolution proceeds, there is some hope to get in a reasonable time a population with "good" individuals. This leads to the following idea:

- Let us develop an evolutionary algorithm which generates populations of PC's. The fitness function should reflect the results of the "Requirement Modelling", that means in this case it should realize a "Fuzzy Multi-Criteria Evaluation" of the PC's.
- As a result we get a finite set of PC's which can be, if needed, ranked by a Fuzzy MADM.

But, unfortunately, there are several problems to apply the concepts of Evolutionary/ Genetic Algorithms to the Design or Configuration Domain. There are very successful applications of those techniques to scheduling type problems, because one can allow inconsistent individuals (schedules) to belong to those populations. For the configuration area it seems to be better to work with populations of consistent individuals (configurations) only. Of course, one could use penalties for infeasibility in the fitness function and use repair algorithms to get a "weakly" inconsistent configuration feasible, but we decided to experiment with the "feasibility" approach.

For the following very short description we assume the basic components of an evolutionary strategy to be known and explain therefore very briefly the crisp part of the approach, while in Section 5.4 the Fuzzy-Fitness evaluation will be explained in detail. The model, the approach and the implementations were elaborated by T. Kriese in his diploma thesis [25].

We will see that it is not very difficult to model the Fuzzy-Design-Multi-Criteria Problem by a fitness function. But it is difficult to find a real good representation of the individuals of the considered population and to deal successfully with the different consistency problems.

In Fig. 34, we give a brief description of the PC-innovative design evolutionary strategy.

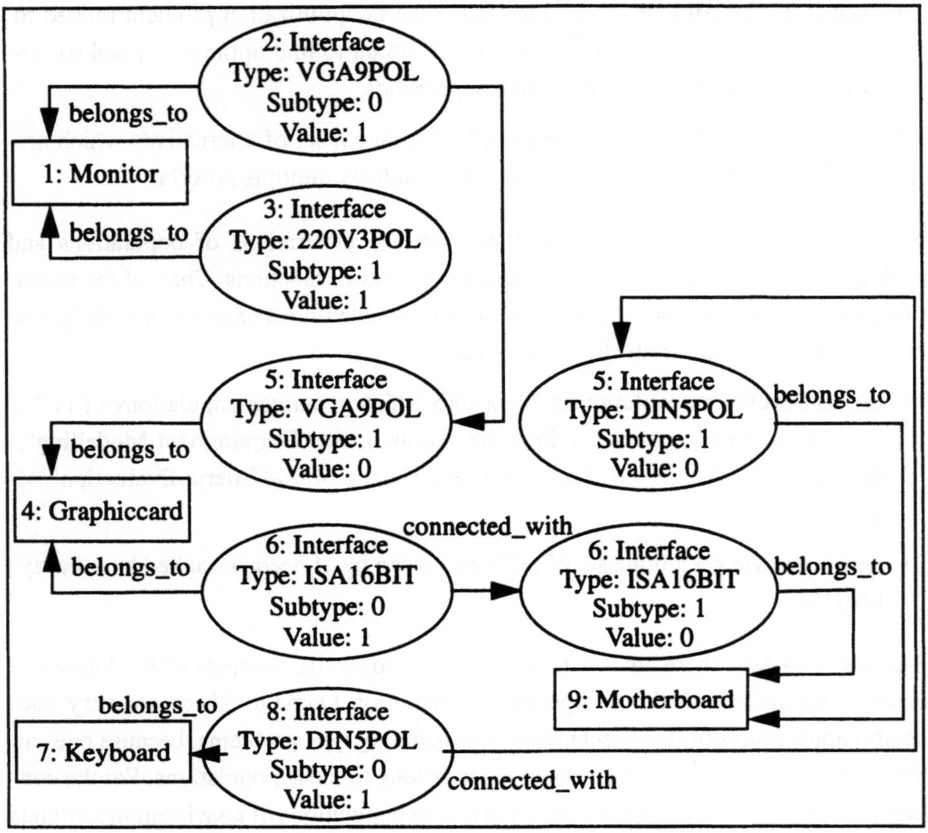

Figure 34: Subgraph to represent a PC

(1) Representation

An individual (a PC) is represented by a directed, marked graph, with:

nodes: physical components, interfaces, building blocks, constraints, (the nodes are attributed), and

edges: belongs_to, connected_with

(2) Consistency

The structural constraints are included into the model of an individual by a graph, that means consistent individuals only are generated by the evolutionary algorithm.

The consistency is guaranteed using a "graph-production-system" (graph-grammar). One particular problem is the use of building blocks.

(3) Mutation and Recombination

There are special operators designed for mutation and recombination, including:

Mutation by:

- exchange of one component by another one,
- delete one component or,
- adding of one component.

If consistency is disturbed after applying such kind of the mutation a second phase of the mutation operator restores consistency, using the productions of the graph grammar.

Recombination by

- generation of a new graph from several given graphs on the level of the building blocks.

(4) Fitness

The fitness function models the precise or imprecise requirements like functional abilities (graphic ability), properties (minimal 300 MB HDD), global constraints (maximal price given), goals (maximal performance) of an individual PC using one real-valued function or a hierarchy of crisp or fuzzy goals.

The details of the model described briefly in steps (1), (2), (3), are best understood by reading [25]. Here we will only concentrate on step (4) in Section 5.4.

3.5 Configuration of a Model in the Environmental Research Area[1]

Modelling is one of the key issues if quantitative (and also qualitative) methods should be applied to problem solving. It is well known that the modelling phase of a decision analysis and decision making approach is the most complicated phase because it requires knowledge of both the domain where the problem is located and the methodology of modelling. In addition, it requires knowledge about solvability of problems.

In the past the research in the area of decision analysis was focused mainly on solution methods, efficient solvers and their implementation. On the other hand, the process of modelling was and is still considered to be an art which should be done by highly specialized experts or groups of experts from different areas. However, there

1. This research was performed within the authors part-time appointment with the International Institute of Applied Systems Analysis (IIASA) in Laxenburg, Austria

are some attempts at least to support the modelling phase by interactive Decision Support Systems (DSS), by Modelling Languages or by Simulation Languages.

In particular the Structured Modelling (SM) introduced by A. Geoffrion [11] and applied successfully to many real world problems is one important approach to the modelling problem.

In our opinion there is a need for a methodology to support the modelling phase of the system and decision analysis, because the research tasks are more and more complex and the models have to describe reality adequately. To get more realistic models, one has to take into account the aspects of uncertainty and imprecision and the multi-criteria-nature of the decision making (optimization) problems.

We start with an object oriented representation and implementation of the system under consideration using the approach described in Sections 2.1 and 2.2. This approach is not restricted to design and configuration in a more narrow sense, but it can be used to design a model of any area of generic or completely specialized objects.

We believe that this kind of modelling supports the maintenance of models, that means, it helps to modify, to generalize and to specialize the model under consideration. Also, it seems that this kind of modelling helps to structure the problem and to think more precisely about objects, their attributes and relationships.

To make the approach more understandable we illustrate it by one of the models in the environmental research area, the RAINS model [2]. Unless some other models like those in [7] would be more appropriate for the application of our approach (because they are more complex) we use the RAINS model for illustration purposes. The RAINS model is "small" enough to understand it quite quickly and also the concepts of imprecision can be illustrated easily by the terms of RAINS.

<u>The Domain Knowledge-Model</u>

In the following we consider our domain knowledge (knowledge about the systems which should be analyzed, designed, simulated or optimized), which consists of three main components

- The world of <u>concepts</u> under consideration, e. g. objects, systems, individuals etc. This world is described by a "is-a-hierarchy" that means a specialization hierarchy with the most general concept, the overall concept, as a root of a respective tree.
- The world of <u>components</u> of those concepts. Components might be also concepts, where examples are again objects, systems etc. The components might be

also represented by is-a-hierarchies that means by specialization hierarchies.

- The relations between concepts and components, concepts and concepts or components and components. For example the concept-components-relation can be represented by "part-of"- or "has-parts"-relations. Also there are general "tupel-relations".

Now we formalize these hierarchies and combine them into one overall hierarchy of concepts, components and relations.

The part-of-hierarchies may be considered as multilevel hierarchies, because a component may also be considered as a system consisting of components of the second level.

It is possible to use only one "conceptual hierarchy" to represent all the hierarchies mentioned above. In Fig. 6 (Section 1) such a conceptual hierarchy is introduced.

We consider the RAINS-Model (Regional Acidification Information and Simulation Model) as it is published in [3]. The model simulates the flow of acidifying pollutants (sulphur and nitrogen species) from so called "Source Regions" to "Environmental Receptors". The kernel of the model is the "Country to Grid Atmospheric Transport Matrix T". Figure 35 shows the Environmental Receptors of Europe (Grids).

For simplicity, we assume the matrix T to be given from external modelling processes.

In [2] the RAINS model is described in detail. To understand the following representation of the model using the framework of the "domain knowledge model", see Fig. 6. The study of [2] is also highly recommended.

The RAINS model considers an Overall-Region-S, the top concept (closed world), for which the analysis and forecast should be performed. In [3] the overall system is specialized immediately to "EUROPE", but it could also be e.g. the USA, South-East-Asia etc.

The components of such Overall System S are "Source-Regions" and "Environmental-Receptors". The Environmental Receptors are land-based receptor sites (shortly called "Grids") where the analysis of deposition of sulphur is made.

The Source-Regions might be specialized to "Countries", "Part-of-Countries" and "Sea-Regions" (ship emissions). Fig. 36 shows the conceptual hierarchy of the RAINS model, where one can easily find the generic objects (also called concepts, abstract entities, classes) as the Overall-Region, Source-Region, Environmental-Recep-

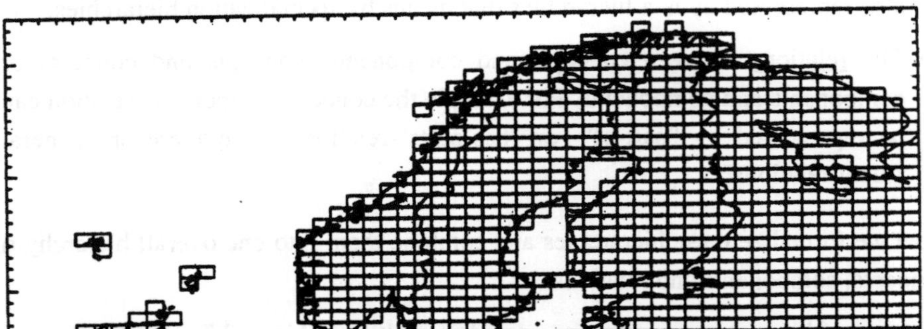

Figure 35: Europe

tor, Country, Part-of-Country, Sea-Region but also the completely specialized concepts, the individuals as e.g. EUROPE, EUROPE-Country-1, South-East-Asia-Country-3, EUROPE-Grid-547.

This conceptual hierarchy, however, does not include all the knowledge needed to do quantitative analysis and decision support. Therefore, quantitative information like "level-of-emission of sulphur", "deposition of sulphur", "cost of reducing emission to a certain level" has to be added to the objects of these hierarchies. We do this, introducing attributes. Generally, if C is any concept of the conceptual hierarchy, characterized by attributes which represent certain properties of C, then we introduce slots (names of the attributes) and fillers which describe the possible slot values.

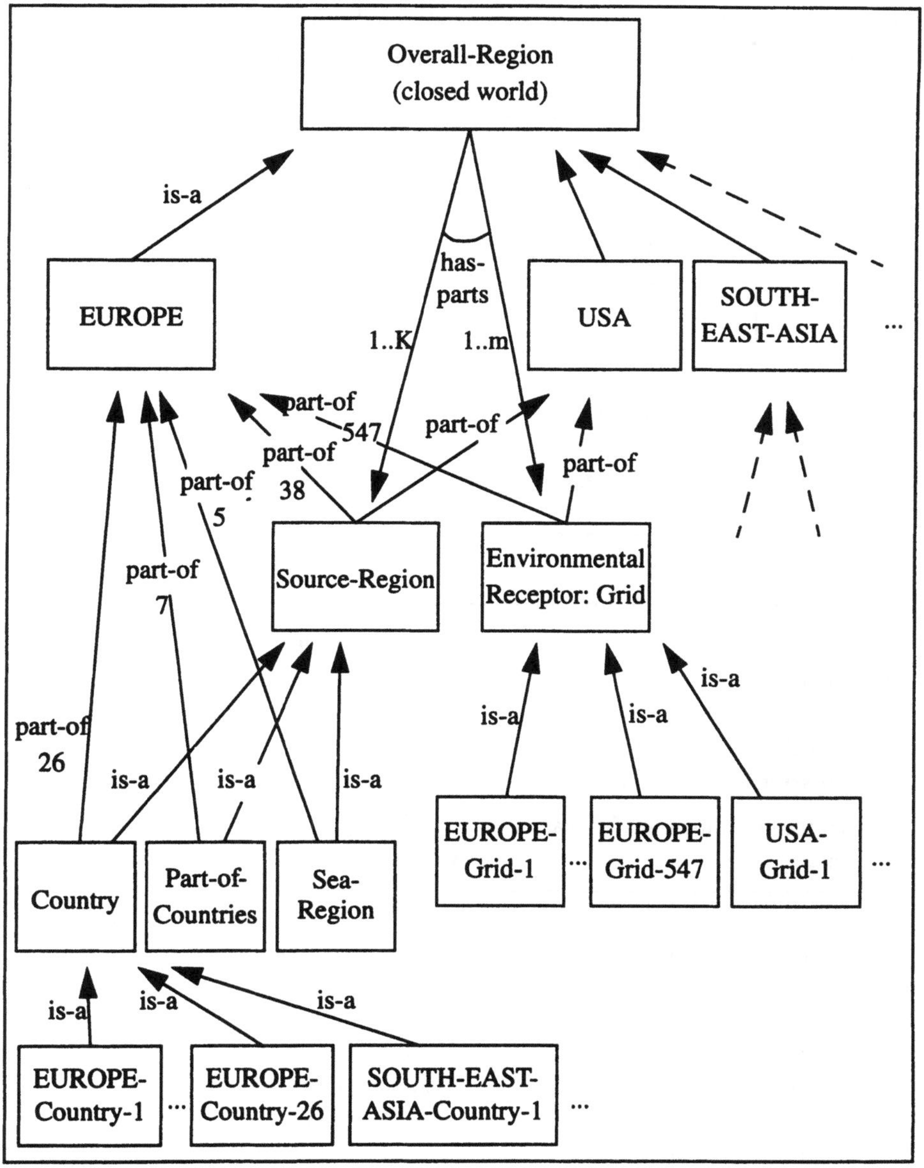

Figure 36: The Conceptual Hierarchy of the RAINS-model
(incomplete because of the complexity)

In Fig. 37 we give an example of a more detailed representation of objects by a FRAME-like description not using a special syntax (which will be, however, introduced in Section 4).

CONCEPT NAME	
Slot-Name1:	Filler
Slot-Name2:	Filler
.	
.	
.	
Slot-NameN:	Filler

Figure 37: Representation of objects by frames

Overall-Concept: S	
is-a:	Modelling-Object
has-parts:	{Source-Region [1, K], Environmental-Receptor [1, m]}
grid-deposition-vector:	$y' = (y_1, ..., y_m)$
emission-level-vector:	$x' = (x_1, ..., x_K)$
source-receptor-transfer-matrix:	$T = (t_{jk})_{m, K}$
cost:	C

Source-Region: E_1	
is-a:	Modelling-Object
is-part-of:	Overall-Concept S
has-subconcepts:	{Country, Part-of-Country, Sea-Region}
level of emission:	x^k
cost-of-reducing-emission-to-level x^k:	$f_k(x^k)$

Figure 38: Representation of hierarchical ordered objects

On the other side Fig. 38 shows how a chain of hierarchically ordered objects is represented (illustrating the inheritance mechanism).

Slot-Name1 gives always a reference to a more general concept (if there is one). Specialization might be realized by restricting the set of possible values (fillers) of at least one of the slots or by adding new slots (Fig. 39).

Environmental-Receptor-Grid: E_2	
is-a:	Modelling-Object
is-part-of:	Overall-Concept S
grid-deposition:	y_j
source-receptor-transfer-coefficients:	$t_{j1}, \ldots, t_{jK}$

Figure 39: Representation of Concepts by Slot-Filler notation to represent the attributes

We remark that a top-node "Modelling-Object" was added to our conceptual hierarchy to identify Overall-Concepts, Source-Region, GRID as specializations of the most abstract object Modelling-Object (Fig. 39a).

EUROPE	
is-a:	Overall-Concept
has-parameters:	K = 38, m = 547

Figure 39a: Concept of EUROPE as an instantiation of Overall-Concept

Using the additional slot "has-parameters" K and m are specified by 38 or 547 respectively. Therefore EUROPE inherits all the slots and slot-values of the Overall-Concept from Fig. 37 by replacing K = 38 and m = 547. Now it depends on the modelling goals, if one wants to add more slots to EUROPE or not. For example the names of countries etc. could be added.

Fig. 39b shows a specialization of the abstract concept Source-Region to an individual EUROPE-Country1.

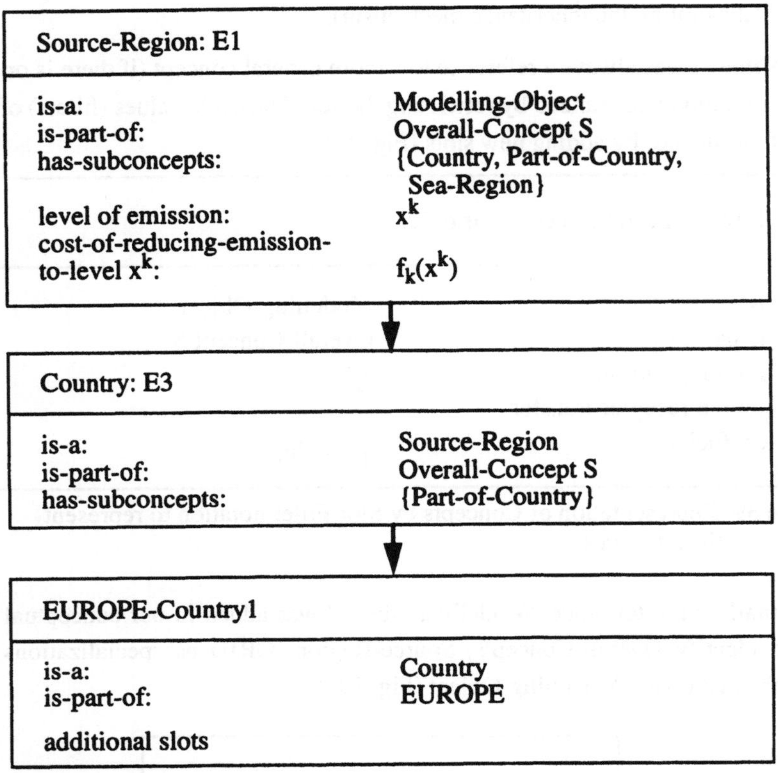

Figure 39b: Example of a specialization chain

EUROPE-Country1 inherits all slots of the more general concepts in the chain (economy of storing information). As an individual it might have additional slots for particular characteristics of this country which might be useful.

Looking at the simple examples shown in these figures one might easily recognize that there are several relations, restrictions which are well known and should be, therefore, added to the knowledge base.

For example:

Cost C of S equals the sum of costs of source-region k over all source-regions.

Deposition at Grid j equals the sum of the product of source-receptor-transfer-coefficient t_{jk} times level of emission x^k of source region k over all source regions.

These constraints (called conceptual constraints, see Section 1) are formulated on the conceptual level. Of course, these constraints must be assigned to the respective concepts of our hierarchy. However, we will not go more into the technical details of implementing the constraints.

It is a big advantage of this kind of modelling that the conceptual constraints are implemented on the level of concept or generic objects by a declarative type of implementation. Therefore, for any particular model say for EUROPE or SOUTH-EAST-ASIA it is not necessary to take care of those constraints, because they are already implemented on the conceptual level and will be instantiated automatically for any particular model-instance.

This principle of object-oriented, declarative way of modelling is one of the big advantages of the approach. It permits models to be generated for particular cases (EUROPE etc.) by instantiation only. In this sense the knowledge model might be viewed as a "core model" (see also similar thoughts by M. Makowski [30]), which defines the domain knowledge only - without considering user requirements. This does not restrict the "feasibility" (the "set of admissible decisions") too much from the beginning. It allows the knowledge base to be changed and maintained relatively easily.

Of course, as the next step, we have to add the modelling of the user requirements to the core model in order to be able to define MCDM or Optimization problems on the basis of the knowledge model.

4 Introduction of KONWERK - a Modular Design and Configuration Tool-Kit

4.1. General Introduction

KONWERK is a tool-kit for conceptual design and configuration of complex systems like, e.g., engineering type or socio-economic systems. Its development was part of the project PROKON[1] [14] which was sponsored by the German Minister of Research and Technology. A central requirement for tools like KONWERK is its applicability in different domains. Thus its concepts must be universal. On the other hand every domain has special properties which should be used to solve problems. Therefore, the modules are divided in basic modules and extension modules which either enhance the abilities of basic modules or provide new abilities.

The basic modules cover the following general tasks in configuration and design systems

- representation of domain objects
- representation and processing of relations, constraints and heuristics
- formulation of the configuration task
- control of the configuration

We will now introduce the modules (basic and extension modules) which together are the kernel of KONWERK. For the support of their implementation we build the following support modules:

- object-oriented enlargement of CommonLispObjectSystem (CLOS),
- definition of interface classes and interface methods between the modules.

The basic modules[2] of KONWERK are necessary in most applications. They offer standard methods for the main tasks in configuration systems. If required by the domain they may be complemented by extension modules. The basic modules of KONWERK are:

- <u>Conceptual hierarchy for the representation of domain object</u>

---

1. This work was supported by the German Federal Ministry of Research and Technology (BMFT) under grant number 01 IW 202 F
2. The basic modules of KONWERK have been implemented by the PROKON-group of Hamburg University

Domain objects and their properties are represented in a conceptual hierarchy which comprises a taxonomic (is-a) and a compositional (part-of) hierarchy. The possible values for the properties are predefined with help of so-called object descriptors.

- Constraints for the representation of relations
  For the representation of configuration restrictions and their processing we use a constraint system. By the use of constraint propagation it is possible to restrict values and to discover inconsistencies.

- Control
  The control is based on a depth-first search and supports chronological back-tracking. The control knowledge is represented in several so called strategies which control the sequence of configuration steps chosen from an agenda.

- Specification of the configuration task
  A configuration task in KONWERK is specified by the selection of a target object which has to be configured. It is possible to select further objects (components of the target object) and to specify their properties (slots).

Extension modules enlarge the functionality of KONWERK in several ways. Due to the concept of a tool-box with a variety of modules the user of KONWERK can build his tool in a very flexible way by selecting the necessary modules. Extension modules extend the basic modules only in a conceptual view. On the implementation level they are often complex modules which contain all the functionality of a basic module and a lot of functional extensions to this basic module.

The conceptual hierarchy may serve as an example for the principle of the modular tool-box. The basis module for representation of domain object is a "simple" version of a taxonomic hierarchy as a basic module. This enables the user to describe his objects in a strict is-a hierarchy and to define their properties. Possible values for these properties are pre-defined by the user. A special property is the compositional relation between objects. This n-ary relation is very important in the configuration process and is especially supported in this module.

There are several extensions to this basic module. We have one extension module for each of them. Thus the knowledge engineer can select the functionality he needs and has not to pay for power he does not want. Extensions are:

- views and mixins
- fuzzy properties

- linguistic values
- spatial properties (2D and 3D)
- fuzzy is-a hierarchy
- measures

In our terms a KONWERK-variant is a collection of modules (basic and extension modules) which is tailor-made for special requirements. This means there is not a single complex KON-WERK-system which solves all configuration problems but there are a lot of KONWERK-variants that are as lean as possible and highly specialized for their type of problem.

In most cases a KONWERK-application consists of a knowledge base and a domain specific user interface. In order to design a KONWERK-application one can:

- select a pre-defined problem specific KONWERK-variant,
- choose all necessary modules and combine them.

A knowledge engineer who wants to implement an application with KONWERK is supported by a description for each module and each variant. These descriptions offer all necessary information about the functionality and map typical application characteristics on modules or variants. Thus the developer can use KONWERK in a way which leads to a solution with all necessary functionality and without any unnecessary complexity.

In the following we briefly describe some basic ideas behind of the extension modules, which enable optimization-based configuration with KONWERK. The respective modules are marked by * in Table 1

Table 1: Survey of the basic and extension modules

task	basic modules	extension modules
• representation of domain objects	• conceptual hierarchy	• views and mixins in the conceptual hierarchy • fuzzy properties in the conceptual hierarchy • linguistic values in the conceptual hierarchy • spatial properties (2D and 3D) • fuzzy is-a hierarchy • measures

task	basic modules	extension modules
• represen-tation and proces-sing of relations	• constraints	• uni-directional functions • rules for the representation of heuristics • constraints with fuzzy variables • spatial relations
• control	• depth-first search with backtracking	• case based configuration • exploration based configuration • interactive configuration • basic optimization module* • optimization module MADM* • linear optimization* • non-linear optimization with constraints* • branch-and-bound optimization* • knowledge based backtracking • knowledge based backtracking with data acceptance • repair instructions • complete search in an explicit solution space • ATMS-based configuration
• task speci-fication	• target object, components and their pro-perties	• requirement modelling (functional task speci-fication, fuzzy task specification, goals and wishes for the optimization)*
• special extensions	•	• integration of simulation methods • compilation of constraints • truth maintenance system • interface to data base systems • interface to CAD systems

4.2 Modelling of the Requirements with KONWERK

In Section 2 we introduced the "Requirement Model". There is an extension module "requirement modelling" of KONWERK which realizes some of the concepts described in Section 2, with other words, which supports the task specification (Table 1). That means, this module helps to formulate the user requirements in both cases, in the crisp case and in the case of imprecise (e.g. linguistic) requirements. There are several possibilities to model such requirements, for example as:

- restrictions
  required components (generic objects) or individuals, required parameter-values, local constraints (binary relations between components), global constraints (restricted resources of the goal system),
- goals
  goals are objectives or criteria for the goal system, which should be maximized or minimized by selecting or constructing an appropriate solution,
- functional requirements
  an user often has requirements formulated as functional abilities of the goal system modelled by crisp or linguistic expressions (e.g. the "car" should be safe and fast, the "computer" should have graphic ability),
- requests
  requests are optional requirements of the types considered above. That means, such requests have a lower priority and might be relaxed or removed during the configuration process.

All the requirements formulated above might be crisp or imprecise. To be able to deal with the imprecise-case, we need extension modules as fuzzy-is-a-hierarchy, compositional hierarchy with fuzzy attributes, fuzzy constraints and goals [18] (see also Section 4.4).

Using a closed-world assumption, all possible local or global constraints, goals, functional requirements, requests have to be formulated by the knowledge engineer and to be represented in the static knowledge base (requirement model). The modelling of the requirements for a determined case (task) by the user (RM) is mainly an interactive process. To this end, there are some requirement-strategies, predefined by the knowledge engineer, which determine a sequence, in which the requirement types are considered and the requirements will be selected or specified by the user. For example, the user will be asked first to define the functional requirements, then to select goals from a predefined list of goals, and then to select constraints, etc. (see Fig. 40).

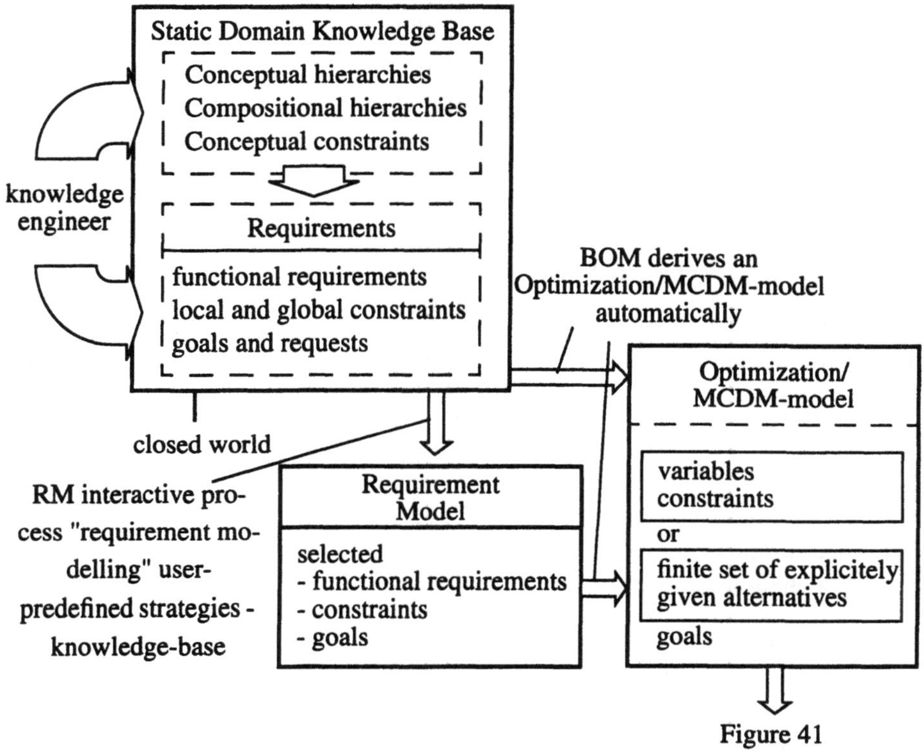

Figure 40: Requirement Modelling and Basic Optimization Module of KONWERK

For a more detailed description of the KONWERK extension-module which supports the requirement-modelling see [32], [47].

It is important to remark that some of the requirements such as goals or constraints might be modelled in different ways, either precise (crisp) or imprecise (fuzzy). To illustrate this decision problem see [43], [51]. For example, the user has to choose his requirement model among a set of different alternatives. Therefore, he starts with an initial model. Then, this model is used for the configuration of the target system. If the result is not acceptable, then the requirement model will be modified. Using an explanation module, the process of modification of the initial requirement model will be supported.

In this chapter we only give a brief introduction of KONWERK. Therefore no details are presented such as the syntax of KONWERK (hierarchies, conceptual constraints, requirements). The syntax is published in [17], [18]. In addition in Sections 4.4 and 5 we will use pieces of the syntax to become more precise in connection with fuzziness in KONWERK and with the five examples we introduced in Section 3.

4.3 Optimization-based Configuration in KONWERK

If the requirement model has been established, we have to specify some of the important components of an optimization model, like the constraints and the goals. The decision variables and several composition-type constraints should be derived from the conceptual hierarchy and the conceptual constraints automatically.

In general, there are several possibilities to formulate an optimization model using the domain knowledge and the requirement model. The task to choose and to represent the optimization model is done in KONWERK by the extension module "Basic Optimization Module - BOM". The BOM acts as an interface between the knowledge-based configuration subsystem and a set of optimization methods and algorithms which are available using optimization-extension modules like "linear optimization", "non-linear optimization with constraints", "optimization module MADM" and others. Fig. 40 and 41 show a simple illustration of the idea.

In Fig. 40 is shown, that the BOM transforms knowledge from the knowledge-base into such a representation of knowledge that is required for an optimization model. In particular, the conceptual constraints that are connected with the conceptual hierarchy and "activated" by the requirement model will be transformed into a representation of constraints that is usually used in the Operations Research.

After this transformation the BOM analyzes the properties of the optimization model and supports the interactive process of the definition of an optimization/MCDM problem related to the model (Fig. 41). For example, if the model consists of N explicitly given design alternatives and there are n goals (so called attributes), then, depending on the information basis a MADM-problem or a Fuzzy-MADM problem would be assigned to the model. Another example could be the following. We have as an optimization model n decision variables, some of which are arbitrary real numbers, some other must be integers. Furthermore there are m linear constraints and r linear goals (objective functions depending linearly on the decision variables). Then, the problem assigned to the model could be a MODM problem (determine the Pareto Set or optimize a linear utility function or apply goal programming).

Summarizing, the BOM does an automatic classification of the model components and supports the definition of an appropriate optimization problem.

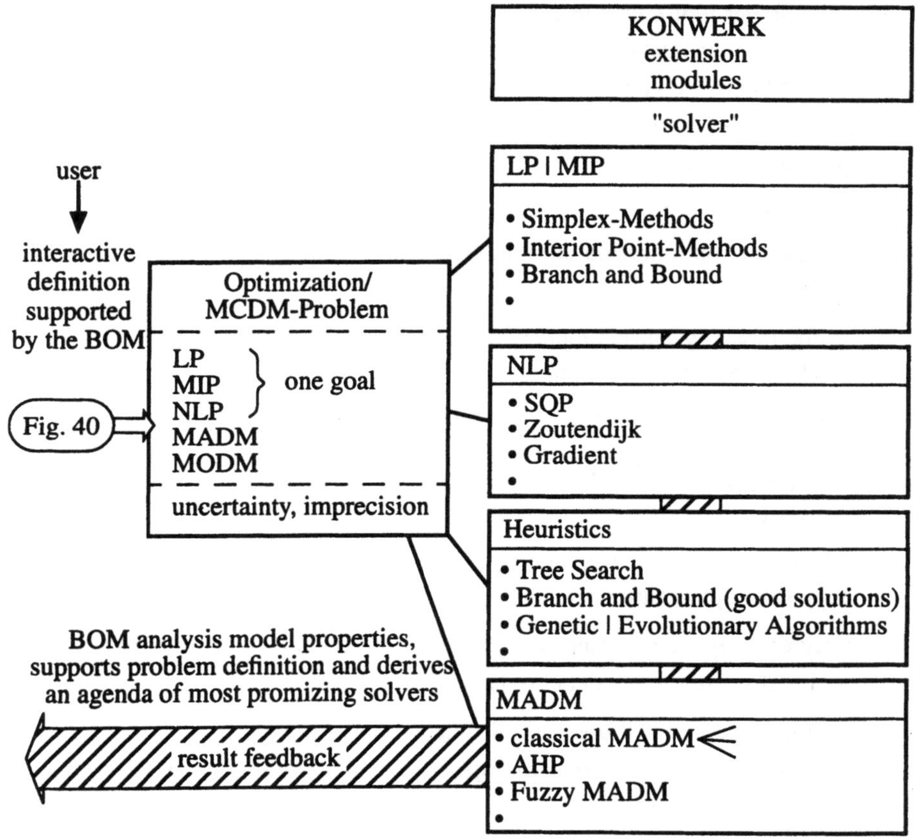

Figure 41: Problems and Solvers

The next task of the BOM is to match this properties with the requirements and the assumptions of the optimization methods which are available. Finally, an agenda of the most appropriate - for the problem - optimization methods is generated and the respective optimization extension module is called to run the chosen optimization algorithm (see Fig. 41).

The results of the solvers are given back to the BOM and the BOM has some strategies predefined to feed these results back to the dynamic or static knowledge base. In particular, if an optimization method is used within the configuration process to perform a configuration step, the result will be used for the "partial construction" of the dynamic knowledge base.

For a more detailed description of the BOM (class Optimization Method, Optimization Strategy), see [8], [9].

Each of the solver-tool boxes in Fig. 41 represents one extension module of KONWERK. The implementation of these modules is still under development. For instance the MADM-Module is almost completely implemented. The conception of each of the modules is such that an additional method can easily be added to the module (open architecture).

The three main possibilities we have in KONWERK to integrate knowledge-based configuration with optimization or multi-criteria decision making are:

1) The overall configuration problem (preliminary design or configuration of the target system) is modelled as an optimization problem. There is no decomposition of the target system.
2) The target system is decomposed into components (subsystems, elements), and the control strategy of the configuration process uses optimization as one particular method to specialize subsystems (or to select subsystems).
3) The target system is decomposed into components, and the control strategy is an optimization strategy e.g., a branch and bound algorithm.

These different ways to combine configuration with optimization are only briefly introduced, but discussed more detail in [18], [21], [46].

In the next sections we will use more detailed KONWERK concepts. An alternative presentation of the declarative way of object-oriented representation of FRAME-type systems is available in [6].

4.4 Fuzziness in KONWERK

In Section 4.1 we only briefly mentioned several possibilities to include fuzziness in KONWERK. Now we will show more fully how imprecise properties of objects might be modelled and processed within KONWERK (see K. Müller [33]). The user is able to

- use a fuzzy set instead of a crisp number while parametrizing an attribute (restricting the slot value of a parameter),
- generate imprecise classes (concepts) on the basis of imprecise properties of the objects,
- solve so called "selection problems" by fuzzy MADM methods
  (in case one has a finite number of explicitly given alternatives and imprecise attributes).

These three possibilities of dealing with fuzziness in KONWERK will be explained in the following using the KONWERK syntax as much as needed.

Imprecise Attributes

The objects defined in the knowledge-base of a KONWERK application have "normally" crisp properties. With the extension "linguistic variables" it is possible to use "linguistic terms" (terms of the linguistic variable) instead of crisp values. This concept is now introduced by a simple example

Example 1

```
(def-do  : name car
         : superconcept domain-object
         : parameters ((performance    [0PS 800PS]        :basicunit "PS")
                       (speed          [0m/h 300m/h]      :outputunit "m/h")
                       (price          [0$ 300000$]       :basicunit "$")
                       (style          [0 100]            :ling-var 'style)
                       (luggage-space  [0.0 1.0]          :ling-var 'size-ls)
                       (safety         [0.0 1.0]          :ling-var 'safety))
         : relations   ((self: nop: optimization-problem 'selection-car)))
```

The syntax used is:
def-do.name car means definition of a domain-object which has the name car. To be able to find this object in an object hierarchy, we declare that car is a domain-object using the slot descriptor *superconcept*.

Also the slot-names performance, speed etc., their ranges and their type descriptions are self-explaining (*PS* means horse power, *m/h* miles per hour, etc.).

Let's now consider the linguistic variable *style* more detailed. It might be defined by three terms: bad, medium, good. In Fig. 42 possible graphical definition of the fuzzy sets "bad", "medium", "good" is represented.

Of course, one has to define this linguistic variable in KONWERK. We do this using the following syntax:

```
(def-linguistic-variable: name style
         : mapping   ((#!bad         {<0 10 1.0 10 40>})
                      (#!medium      {<10 40 1.0 60 90>})
                      (#!good        {<60 90 1.0 90 100>})))
```

Mapping describes the three membership-functions in Fig. 42 which define the linguistic terms. The user of KONWERK can choose for the variable *style* either any value

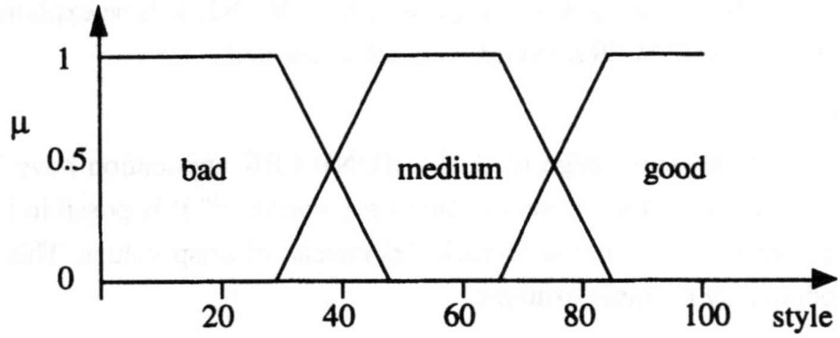

Figure 42: Linguistic variable "Style" (of object car)

in [0 100], a subinterval of this interval, or one of the linguistic values *bad, medium, good.*

Imprecise Object Classes (Generic Objects, Concepts)

In Section 1 (Fig. 8 - 10) we explained imprecise object classes and the process of specialization on a descriptive level. Now we explain how this specialization process is implemented in KONWERK. Using KONWERK's basic modules, specialization is possible only in the crisp case. If we use linguistic variables to characterize attributes, then we can use α-cuts to describe the specialization.

Example 2

Let's consider again a toy-example from the "car-configuration" domain. We assume we have 6 cars each characterized by the size-of-engine (in cm^3) only (see Table 2). Each car-instance has a crisp value of the parameter *size-of-engine*. Now, we can declare the domain object car in the knowledge base by

```
(def-do  : name car
         : superconcept domain-object
         : parameter  ((size-of-engine     [0.0 3.5]:        ling-var 'size-engine)
                        (alpha-cut          [0.0 1.0])))
```

This means the parameter *size-of-engine* of the object *car* is associated with the linguistic variable *size-engine* (Fig. 42a). Now we are able to define this linguistic variable by

```
(def-linguistic-variable: name size-engine
         : mapping    ((#!small          {<0.0 0.0 1.0 1.5 2.0>})
                        (#!medium         {<1.5 1.75 1.0 2.25 2.5>})
                        (#!big            {<2.0 2.5 1.0 3.5 3.5>})))
```

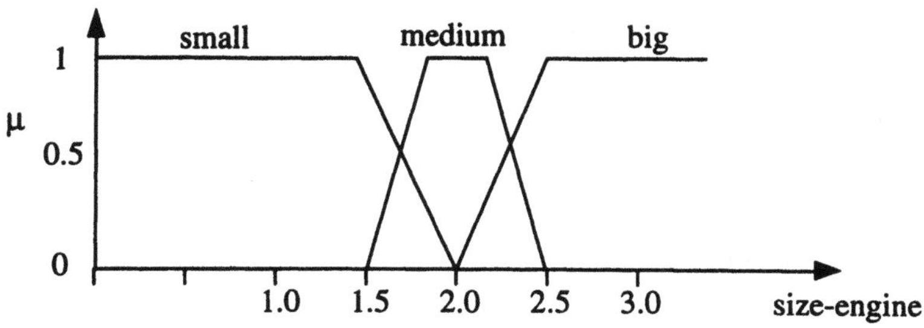

Figure 42a: Linguistic variable *size-engine* (of object car)

In this simple example the generic object *car* has only one parameter which is described by a linguistic variable. Of course (see example 1) it is possible to have several such kind of parameters.

Table 2[1]: A small car database

def-do: name Fiat-Tipo-2.0-16V :superconcept car :parameter ((size-of-engine 1.995))	def-do: name Alfa-Romeo-SZ :superconcept car :parameter ((size-of-engine 2.959))	def-do: name Alfa-Romeo-33-16V :superconcept car :parameter ((size-of-engine 1.712))
def-do: name Lancia-Stratos :superconcept car :parameter ((size-of-engine 2.418))	def-do: name Maserati-222-SE :superconcept car :parameter ((size-of-engine 2.790))	def-do: name Alfa-Romeo-Spider :superconcept car :parameter ((size-of-engine 1.962))

Assume the user asks for a car using the linguistic term *#!medium* for *size-of-engine*. Then for each element in the database (each of the 6 cars) the membership value for the fuzzy set *#!medium* will be determined using the definition of the linguistic variable. The membership values are shown in Table 3.

As a default α-cut the value 1.0 is predefined. Therefore, we would get as possible specializations:

Fiat-Tipo-2.0-16V, Alfa-Romeo-Spider

1. In this section we use car-names to be more illustrative. However, the data assigned to those names is not necessary equal to the true data.

Table 3: Membership values

cars-name	size-of-engine	$\mu_{medium, size-of-engine}$
Fiat-Tipo-2.0-16V	1.995	1.0
Alfa-Romeo-SZ	2.959	0.0
Alfa-Romeo-33-16V	1.712	0.848
Lancia-Stratos	2.418	0.328
Maserati-222-SE	2.790	0.0
Alfa-Romeo-Spider	1.962	1.0

The user can change the α-cut, for example let's take 0.3. Then possible specializations of #!medium are

Fiat-Tipo-2.0-16V
Alfa-Romeo-33-16V
Lancia-Stratos
Alfa-Romeo-Spider

We started with a class of crisp objects, but one of the attributes of the objects is described imprecisely.

Of course, the more interesting case is the case of more than one linguistic variable associated to an object. In this case one has to use the Fuzzy-AND which is defined in KONWERK by default as the Minimum operator. But in addition any other operator might be used.

<u>Fuzzy MADM</u>

Fuzzy MADM is a module of KONWERK which solves ranking or screening problems over finite sets of explicitly given alternatives if the alternatives are described by imprecise attributes. It is not the aim of this paragraph to describe Fuzzy MADM methods (see [5]), but we will again illustrate the approach which is available in KONWERK using a simple example.

<u>Example 3</u>

Again the task should be to consider 6 cars and in addition, now, to select one of them. There is a database with special cars and their properties:

Table 4: A small example database

DATABASE	size-of-engine [cm³]	cy-lin-der	perfor-mance [PS]	max-speed [km/h]	price [DM]
Fiat-Tipo-2.0-16V	1.995	4	146	212	31700
Alfa-Romeo-SZ	2.959	6	210	245	100000
Alfa-Romeo-33-16V	1.712	4	132	205	28000
Lancia-Stratos	2.418	6	192	230	100000
Maserati-222-SE	2.790	6	225	230	179800
Alfa-Romeo-Spider	1.962	4	125	193	40500

The data (in particular the prices) in Table 4 are not real, but chosen artificially.

The idea is now that in addition to the parameters in Table 4 there are parameters defined in the domain object car, which are linguistic variables <u>and</u> have to be user defined. In this example we choose *luggage-space* and *safety* to be parameters of this type. Then we have to define the problem using KONWERK syntax:

```
(def-do  : name car
         : superconcept domain-object
         : parameters ((size-of-engine   [0cm³ 3.5cm³]   :ling-var 'size-engine)
                       (performance      [0PS 800PS]     :basicunit "PS")
                       (speed            [0km/h 400km/h] :outputunit "km/h")
                       (price            [0DM 400000DM]:basicunit "DM")
                       (style            [0 100]         :ling-var 'style)
                       (luggage-space    [0.0 1.0]       :ling-var 'size-ls)
                       (safety           [0.0 1.0]       :ling-var 'car-safety))
         : relations   ((self: nop: optimization-problem 'selection-car)))
```

The two linguistic variables for which there is no information in the data-base are defined by:

```
(def-linguistic-variable: name size-ls
         : adjective    (small big))
```

(def-linguistic-variable: name car-safety
: adjective (bad good))

In addition there must be a selection problem defined for the cars by a declaration in the knowledge base:

(def-selection-problem: name selection-car
: goals (safe big-luggage-space small-price high-speed)
: object car
: penalty-for-a-bad-evaluation: no
: compensation: yes
: ranking alternatives: yes)

The definition of a problem such as the one above means, apply a ranking rather than a screening method, use compensation for bad evaluations and not penalties in addition. These requirements lead to the conclusion to apply "Simple Additive Weighting Method".

In addition the 4 goals defined in the *goals* slot must be declared.

(def-goal : name safe
: slot safety
: evaluation nil
: direction max
: reservation-level-conj ((car 0.5))
: reservation-level-disj ((car 0.5))
: goal-value ((car 0.5)))

(def-goal : name big-luggage-space
: slot luggage-space
: evaluation nil
: direction max
: reservation-level-conj ((car 0.5))
: reservation-level-disj ((car 0.5))
: goal-value ((car 0.5)))

(def-goal : name small-price
: slot price
: evaluation nil
: direction min

: *reservation-level-conj ((car 50000DM))*
: *reservation-level-disj ((car 100000DM))*
: *goal-value ((car 25000DM)))*

(def-goal : name high-speed
: *slot max-speed*
: *evaluation nil*
: *direction max*
: *reservation-level-conj ((car 180km/h))*
: *reservation-level-disj ((car 300km/h))*
: *goal-value ((car 280km/h)))*

The goals make reference to the attributes to which they are connected first (e. g., goal with name safe refers to the slot safety). Evaluation nil means that there are not any known predefined evaluations of the considered goal. Also, there are some levels that should be at least reached (*reservation-level-conj, reservation-level-disj*) or the distance to those values should be minimized (*goal-value* desired). Assume during a configuration process that a selection-problem is new on the agenda. The user will be asked to give his linguistic evaluation of size-of-luggage-space and safety using the predefined terms (all other data are known for each car from the database in Table 4). The terms bad-good, small-big are now taken as opponents and following Chen and Hwangs method [5] the user has the possibility to choose at first one of the 8 scales of Chen and Hwang and than to select his linguistic evaluation from the scales.
Fig. 43 illustrates the scales using the example bad-good.

The user chooses the scale and on the selected scale the desired linguistic value. For example for the first car in the database and the safety-evaluation the user could choose scale 6 and give the evaluation safety-of-the-car by "more-or-less-good". Then, Chen and Hwangs method uses left-and-right score method for Defuzzification and performs a ranking in a second aggregation phase of all attributes [5]. We explained this Fuzzy MADM approach implemented in KONWERK as an example here because this method is widely used in practical applications (see Section 5.5) and [18].

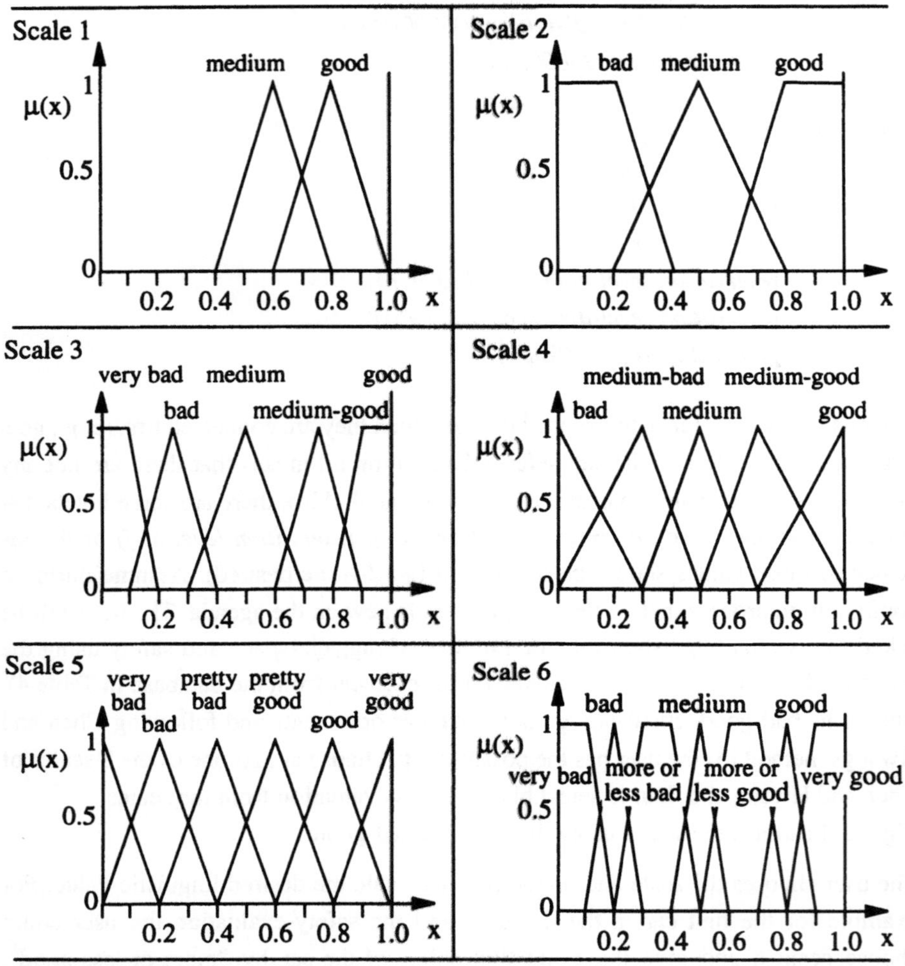

Scale 7 and Scale 8 see [5].

Figure 43: Chen and Hwang's scales to represent imprecise linguistic terms

5 Solving Applications Using the Integrated Approach of Knowledge-Based Techniques with Fuzzy Logic and Fuzzy MCDM

In this section we consider the real world applications defined in Section 3 and we describe the solution concepts by using the methodology introduced in Section 2 and partly the toolkit KONWERK introduced in Section 4. We will only show typical steps of the solution processes rather than the completely computed examples because of the restricted space available. For the complete description of the examples we will provide the reader with appropriate references.

5.1 Passenger Aircraft Cabin Layout Application

5.1.1 Fuzzy Modelling for the Aircraft Cabin Layout Problem

In this Section we describe briefly some possibilities to define fuzzy goals for the cabin layout problem [23]. There are mainly four typical fuzzy goals related to this task:

(1) The distance between two seat-rows ("seat-pitch") should be "comfortable" for the passengers. Of course, there are several alternatives to model "comfortable distance", because comfortable depends on the length of the flights (short- or long-distance), on what people are expecting (business or tourist class) and on individual characteristics of the passengers. However, it is possible to define a fuzzy set "Comfortable-Distance" = CD by introducing a membership-function $\mu_{CD}(d)$ where d is the distance-measure between two rows.

Using this measure for the distance between two neighbors rows, we are able to introduce an overall measure for one class of the airplane taking into account that row-distances could be different.

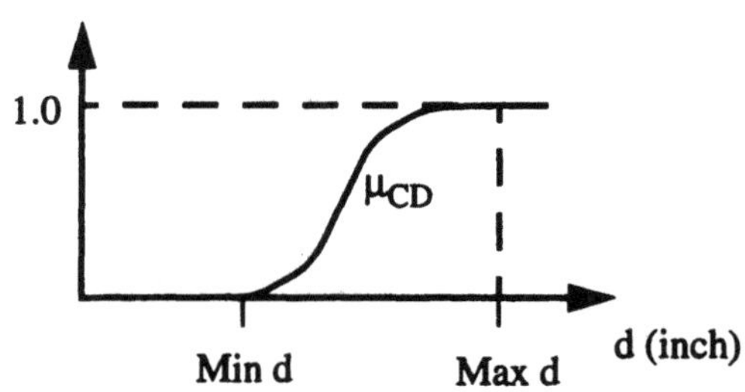

Figure 44a: Fuzzy Set "Comfortable-Distance"

(2) The possible number of seats in a class depends on the "distance between rows" and on the other layout decisions (lavatories etc.). It could be also considered a fuzzy goal as follows:

There is a fuzzy set "Large-Seat-Number" = LSN defined over the set of integers, context dependent (on the airplane-type, class etc.) defined by a membership-function $\mu_{LSN}(n)$, see Fig. 44b.

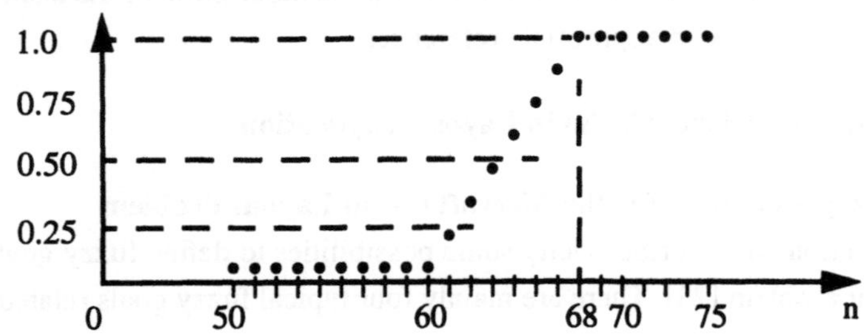

Figure 44b: Fuzzy Set "Large-Seat-Number" related to a business class

(3) Another fuzzy goal is defined by the delivery-time for cabin interior components which might differ from nine months to two years.

We introduce the fuzzy set "Fast-Delivery-Time" = FDT for an individual cabin interior component, e. g., for a galley, by a membership-function $\mu_{FDT}(t)$, see Fig. 44c.

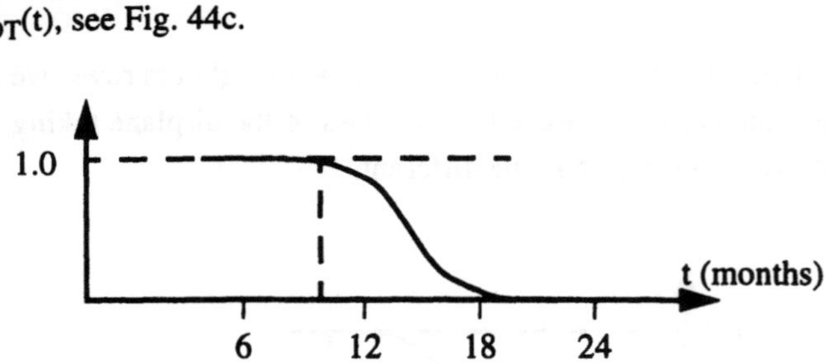

Figure 44c: Fuzzy-Set "Fast Delivery Time"

Each cabin interior component has a so called "lead-time" that describes the time needed from order to delivery depending on the supplier of that component. This means the "Fast-Delivery-Time" fuzzy sets must be carefully defined for each interior component.

(4) The price for a cabin layout is defined by the sum of the prices of all its cabin interior components. The price for a single component differs depending on its place in the cabin. But, instead of a crisp goal for the whole layout, we can ask for a "low-price-layout", using a respective fuzzy set, to define what "low-price" means.

Using these four Fuzzy-Goals, there are several alternatives to model "the optimality" of a cabin layout. We need an aggregation operator to define an overall criterion (top-goal). Again there are several alternatives. We use compensatory operators to compensate "Comfortable-Distance" (passengers appreciate this goal) with "Large-Seat-Number" (airlines goal), and "Low-Price-Layout" with "Fast-Delivery".

Therefore, for a specified cabin-layout we are able to compute a degree of overall-satisfiability, and (if we are able to generate several layouts) to find a ranking where the top layout is the optimal-one according to the finite set of layout alternatives.

In Section 5.3. we will briefly show how we can generate such a finite set of alternatives using KONWERK's "Branch and Bound" module and how we rank the alternatives using the Fuzzy-MADM-module of KONWERK. But, before doing this we will briefly illustrate how the knowledge base can be modelled using KONWERK.

5.1.2. The Knowledge Model: Conceptual Hierarchy, Measures, Viewpoints & Multiple Inheritance, 2D-Space, Constraints, Formulas and Control

The module *Conceptual Hierarchy* of KONWERK allows objects to be described in a very efficient manner. The following two examples in KONWERK-syntax demonstrate the possibilities of the modules *Conceptual Hierarchy, Measures, Viewpoints & Multiple Inheritance* and *2D-Space* for the representation of objects.

```
(def-do :name cabin-interior-component
        :superconcept  (domain-object rectangle)
        :parameter     ((ref-nr [integer 2500000 2599999])
                        (x-range [float 404inch 2408inch])
                        (y-range [float -105inch 105inch])
                        (length [float 1inch 200inch])
                        (width [float 1inch 200inch])))
```

The representation of the object *cabin-interior-component* demonstrates the is-a hierarchy of the basic module *Conceptual Hierarchy.* By the use of the extension module *Views & Multiple Inheritance* it is possible to inherit properties not only from the object *domain-object* but also from *rectangle* which is defined in the extension module *2D-Space*. The extension module *Measures* allows the usage of the measure inch in the slot length.

```
(def-do :name galley
        :superconcept cabin-interior-component
        :relations     ((galley-in (a passenger-class)))
        :parameter     ((ref-nr [integer 2531000 2533999])
                       (door {1 2 4})
                       (trolleys {0 2 3 4 5 6 7 8 9 10})
                       (half-size-trolleys {0 1 2 3 4 5})
                       (meals [integer 28 140])
                       (type {longitudinal transversal})
                       (height {full half} (default 'full))))
```

The object *galley* inherits all properties of *cabin-interior-component*. In order to specialize the *cabin-interior-component* to a *galley* some of its slots are restricted (e. g., *ref-nr*) and some new slots are added.

Restrictions between objects are expressed in a special constraint language. The following example demonstrates the use of the basic module Constraints and the extension module *Formula*.

```
(def-conceptual-constraint
        :name max-number-of-seats-between-door-2-and-3
        :variable-pattern-pairs ((?s :all
                       :concept bench
                       :parameter ((xpos [963.41 1562]))))
        :formulas ("sum (?s.capacity) < 155"))
```

This constraint describes a certification rule that limits the number of seats between the doors 2 and 3 to 155. This certification rule is obeyed by identifying all benches between the doors (defined by their x-position) and adding their capacities (a bench may contain 1, 2, or 3 seats). Thus it is impossible to place more than 155 seats between the two doors.

The following conceptual constraint ensures the customers demand for a certain seat pitch. It does not matter whether a seat has to be placed in the first class with a seat pitch of 40 inches or in the tourist class with a seat pitch of 30 inches. This single constraint works for all seats in all classes. It uses a constraint relation that is defined in the extension module *2D-Space*.

```
(def-conceptual-constraint :name seat-pitch
        :variable-pattern-pairs
          ((?s1 :concept seat)
          (?s2 :concept seat
          :parameter ((tg-left-neighbor ?s1))))
        :constraint-calls ( (to-difference-frontside ?s1 ?s2 (?s2 pitch))))
```

For more detailed explanation of the conceptual constraints in KONWERK please refer to [18]. For a more complete description of the knowledge base of the aircraft cabin layout problem see [22]. Before we start to explain the optimization-based configuration using fuzzy goals, we will explain the "normal" knowledge-based approach to solve the problem by defining so called control strategies.

The following strategy for the placement of cabin attendant seats gives an impression of how the basic module *Control* is used in XKL [22], [23].

```
(def-strategy :name place-cas
        :strategy-class depth-first-strategy
        :priority 15
        :configuration-step-focus focus-cas
        :agenda-criteria (prefer-specialize
                        prefer-xpos
                        prefer-parametrization)
        :proceeding (compute take-default ask-user))
```

The strategy uses depth-first search, and has a priority that determines whether it will be chosen when several strategies can be applied. It uses a focus to filter the agenda. The filtered configuration steps are processed in an order that determines that as a first step the cabin attendant seat should be specialized. The system should determine the x-position of the cabin attendant seat as a second step before all the other parameters are determined. The system proceeds by first trying to compute a value then looking for a default value and as a last resort asking the user for a value.

One might see that it is possible and useful to fuzzify strategies like the one considered above in order to generate several different solutions.

In this Section we gave a short description of the conventional 'configuration' approach. We described the knowledge representation and gave an impression of the proceeding. A configuration is built step by step. The system ensures consistency for each step but does not consider any optimization criteria. However, the knowledge-base with the conceptual constraints briefly mentioned above is also needed and useful for optimization. Now we would like to introduce methods that enable us to 'optimize' our solution.

5.1.3. An Example for the Use of the Optimization Module

Let us consider a 2-class-layout with a business and a tourist class. In our example (Table 5) the customer wants exactly 68 seats in the business class and as many seats as possible in the tourist class.

Table 5: Requirements for a Cabin Layout

Customer's Requirements for	Business Class	Tourist Class
Passengers (Seats)	68	max
Seat Pitch	35	30
Seats Abreast	6	9
Passengers per Lavatory	35	45
Passengers per Cabin Attendant	25	50
Meals per Passenger	2	1

Thus we have from the initialy formulated requirements two optimization tasks:

- minimization of the needed space for the business class
- maximization of the seats in the tourist class

The problem of optimizing the whole cabin is quite hard for an object oriented sequential configuration system. Nearly 300 objects have to be considered. The complexity of this problem is reduced by decomposing the cabin into its classes. The business class in our example consists of only about 70 objects. Due to the large number of possibilities for the placement of an item (a seat can be placed at nearly 6000 positions) this is a considerable reduction. The trade off of this decomposition approach is that the result consists of optimal and consistent classes which are composed into a consistent cabin which may not be optimal.

In the following, we consider the layout of the business class as an example only.

Our approach is the following:

Phase 1. We use the Branch and Bound-module of KONWERK to generate so called "good solutions". This module was elaborated by S. Järvinen [21] and the underlying algorithms described in Section 5.

Phase 2. The finite set of good solutions (the finite set of layouts for the business-class) will be considered as set of alternatives for the Fuzzy-MADM module of KONWERK (Section 4.4). The fuzzy goals that are needed for the Fuzzy-MADM we introduced in Section 5.1.1. As a result, we get a ranking of the "good-layouts" generated by the Branch and Bound algorithm.

Phase 1

It is impossible to go into the details of the BAB-approach, however, in the following we will give a brief outline of the method.

To define, what "good solution" means, we define three crisp goals:

- minimization of the space needed NS_{bc} for the business class,
- minimization of the price PIC_{bc} for the cabin interior components of the business class,
- minimization of the delivery time DT_{bc} for the components in the business class.

We propose that, if the layout is determined, we are able to measure each of the goals above by a real number denoted again by NS_{bc}, PIC_{bc}, DT_{bc}. Now, we introduce reservation-levels as upper bounds for each of these criteria to define what "good" means. A good layout fulfills each of these constraints:

$$\left.\begin{array}{lccc} & NS_{bc} & \leq & \overline{NS_{bc}} \\ \text{and} & PIC_{bc} & \leq & \overline{PIC_{bc}} \\ \text{and} & DT_{bc} & \leq & \overline{DT_{bc}} \end{array}\right\} \quad (x)$$

where $\overline{NS_{bc}}$, $\overline{PIC_{bc}}$, $\overline{DT_{bc}}$ are the upper bounds.

The BAB algorithm now realizes a tree search strategy combined with the bounding. It is assumed (see Fig. 45) that the configuration process is a sequential process starting either with selecting one of the possible lavatories and positioning the selected one or with the respective step for galleys. It is also assumed that on the basis of such an incomplete layout we are able to estimates of the criteria NS_{bc}, PIC_{bc}, DT_{bc}, more precisely we calculate upper bounds for the criteria. These upper bounds $\overline{NS_{bc}}$, $\overline{PIC_{bc}}$, $\overline{DT_{bc}}$ give us the possibility to terminate an incomplete construction if it is identified to be not a good one.

Finally, this kind of "best-first-search" combined with bounding terminates if a given number of good solutions is reached, or if only a smaller set of good solutions (which might be empty) can be found. If the set is empty one needs to decrease the bounds or to relax the constraints and to consider them as fuzzy constraints.

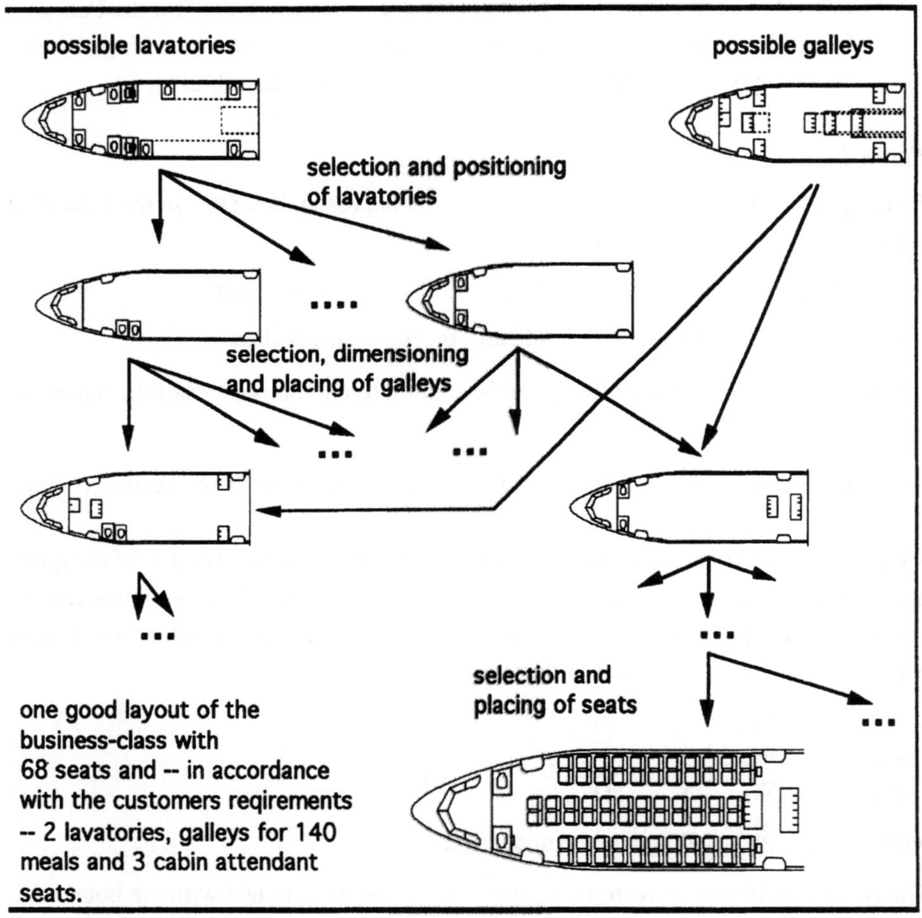

Figure 45: Illustration of the Best-First-Search - Part of the BaB Algorithm

Each of the good-solutions generated by the BAB algorithm is a feasible solution because KONWERK's knowledge base with the conceptual constraints ensures this feasibility by Constraint-Propagation at each of the stages of the algorithm illustrated in Fig. 45.

Phase 2

In Phase 2 each of the four fuzzy goals of Section 5.1.1 has to be modelled by KONWERK as described in Section 4.4. Then, the simplest approach is to use the known crisp data for each of the layouts; to find the respective membership values from the definition of the fuzzy goals, and to determine an aggregated value for each of the layouts. This gives us a ranking of all the good layouts. However, there is also another possiblity.

We have a graphic representation of the different layouts and we can add the crisp data from Phase 1. Then, without using membership functions for the fuzzy goals in Section 5.1.1 we can just linguistically define goals:

- comfortable-distance: with adjectives (bad, good)
- large-seat-number: (small, big)
- fast-delivery-time: (slow, quick)
- low-price-layout:(cheap, expensive)

Then we can show each of the layouts (good layouts) to the user and ask him to give the evaluation using Chen and Hwangs 8 scales. After this subjective evaluation KONWERK generates a ranking of the "good layouts" according to the evaluation with respect to the four fuzzy goals.

5.2. Preliminary Design and Rational Evaluation of Future Space Launch Systems

In Section 3.2. we defined the preliminary design and configuration task and we introduced a conceptual hierarchy for space launch systems. Now, we will consider several implementation oriented issues. We illustrate the domain knowledge model by showing selected concepts of the conceptual hierarchy represented using the KONWERK syntax. The first declarations define STS (Space Transportation System). The parameters are defined either by a finite set (propulsion-type is either rocket or airbreather) or by intervals of real numbers e.g. (take-off-mass is a real number in the interval from 0.0 to infinity).

The most important is the modelling of the has-parts relation (see Fig. 6 in Section 1). It realizes a compositional hierarchy with arc-quantifiers, e.g., there are 1 to 9 STS-parts, 1 to 5 of them are Vehicle-Components etc.

```
(def-do :conceptname STS
        :superconcept domainobject
        :parameter     ((propulsion-type {rocket airbreather})
                       ...
                       (take-off-mass [0.0 inf])
                       (length [0.0 inf])
                       (diameter [0.0 inf]))
        :relations     ((has-parts { # [(a STS-PART) 1 9] :=
                       # [(a PAYLOAD) 0 1]
                       # [(a VEHICLE-COMPONENT) 1 5]
                       # [(a AUXILIARY-COMPONENT) 0 3]}))))
```

A first specialization is the ROCKET-SYSTEM concept, where the parameter *propulsion type* is determined, the *take-off-mass* is restricted to an intervall of real numbers, and a new parameter, the *number-of-stages* is restricted to set of three linguistical values:

```
(def-do :conceptname ROCKET-SYSTEM
        :superconcept STS
        :parameter     ((propulsion-type 'rocket)
                       (number-of-stages {1-stage 2-stages 3-stages})
                       ...
                       (take-off-mass [0.0 50]))
        :relations     ((has-parts {# [(a STS-PART) 1 7] :=
                       # [(a PAYLOAD) 0 1]
                       # [(a BOOSTER) 0 2]
                       # [(a STAGE) 1 3]
                       # [(a AUXILIARY-COMPONENT) 0 3]}))))
```

Furthermore, the parts of a Rocket System are specialized and restricted too.

Another important part of the Knowledge Model is the definition of the conceptual constraints. Constraints for example:

> "The tank volume of the solid-propellant is equal to the solid-propellant mass times the reciporcal value of the solid-propellant density"

are "conceptual", because they define relations on the conceptual level. Therefore, declaring them together with the concepts results in automatically consistent solutions of the specialization process. With other words, there is no need to require those constraints for each specialization, if they are once declared on the conceptual level.

The example which follows declares the conceptual constraint which computes the take-off-mass of an STS by addition of two so-called sums with defined components (formula-part of the constraint). Importantly is that it is easy to see how the formula-part of the constraint is completed with slots of objects of the hierarchy by using variable-pattern-pairs.

```
(def-conceptual-constraint
        :name STS-TAKE-OFF-MASS
        :variable-pattern-pairs
                    ((?STS: name STS)
                    (?VC: all: name Vehicle-component
                    :relations ((part-of ?STS)))
                    (?P: all: name Payload
                    :relations ((part-of ?STS)))
        :formula    ("?STS.take-off-mass =
                    sum-with-parts(?VC.m_0,
                    ?STS.has-parts, Vehicle-Component)
                    + sum-with-parts(?P.capacity,
                    ?STS.has-parts, Payload)"))
```

Unfortunately, all of the details do not become completely clear when examining to this example of a conceptual constraint. However, to make it completely clear, we would have to declare more objects of the domain and we would have to define the syntax precisely. For readers interested in the details we refer to [13], [18], [23]. The whole domain knowledge-base is implemented by T. Goudarzi-Pour [13].
Next, the Preliminary Design Phase Algorithm will be briefly illustrated.

Preliminary Design Phase Algorithm

Input of the algorithm is the payload mass to bring to a prescribed orbit. The algorithm performs an estimation of a start mass by making the basic design decisions (e.g., technology level, Δv requirement, number of stages, number of booster types, etc.). The algorithm operates on the whole knowledge base explained above and uses in addition expert knowledge for the control flow. The algorithm is interactive rather than complete automatic. In Fig. 46 the algorithm, restricted to the case of "Rocket Systems", is briefly presented. For a more detailed explanation see [13].

The number of STSs which one could get from the algorithm is combinatorially restricted to several hundred. However, for a given mission (including the payload

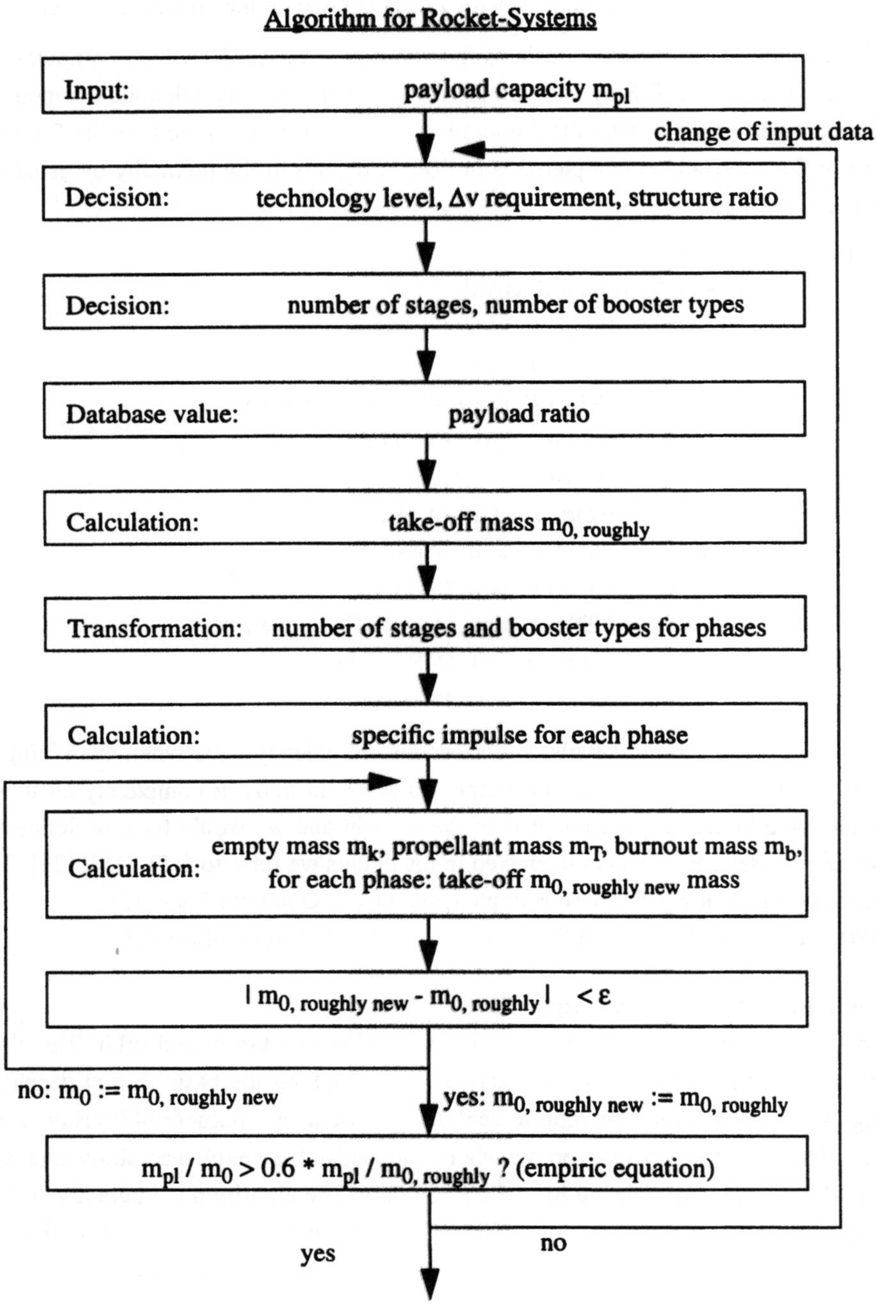

Figure 46: An algorithm to estimate the start mass

capacity m_{pl}) there might only be several tens of systems feasible (able to fulfill the mission). Therefore after this preliminary design phase (where a finite set of STS is selected, which is characterized by the ability to fulfill the given mission) there is a second phase where several other decisions will be made and computations will be performed in order to describe each of the variants (generic STS) more precisely. This second stage decision and computations are very important in order to be able to make a multi-criteria evaluation and possibly ranking of the resulting STS.

Assuming the second phase computations are made (for details see [13]), then we can apply Saaty's AHP method to rank the alternatives for STS fulfilling the mission.

Saaty's AHP for rational evaluation and selection of STS

First of all one has to define a goal hierarchy. It is not easy to find a commonly accepted hierarchy of goals, subgoals etc. In the following, we use the:

first-level goals:

- Transportation Performance (TP)
- Economy (E)
- Environmental Compatibility (EC)
- Availability (A)
- Safety (S)
- Necessity of Space-Transportation (N)
- Access to Space (AS)
- National/European Prestige (NEP)

Using this kind of goals makes the decision process visible and repeatable. Of course the relevance of goals is task (mission) dependent.

At a second level we have the subgoals of

TP:	- Mass
Transportation	- Volume
Performance	- Acceleration
	- Vibration
	- Precision of desired orbit
	- Multiple-Start-Ability
	- Transportation of humans
	- Reentry Ability
E:	- Specific Transportation Costs
Economy	- Start Costs SC

	- Market Position
	- Duration of development
	- Operational Time
	- Reliablity
	- Technology Level
	- Delays (respect to start date)
	- Preparation Time
EC	- Impact of Exhaust Gases (EG)
Environmental	- Noise Duration (ND)
Compatibility	- Noise Intensity (NI)
	- Space Debris (SD)
	- Infrastructure (IS)
	- Impact of Reentering or exploding parts (P)
	- Production of vehicles and propellant (PRO)
A	- Delay in Preparation
Availability	
S	- Reliablity
Safety	- Rescue System

In addition there are subgoals of the StartCosts:

- Development Costs
- Production Costs
- Financing Costs
- Operating Costs

Fig. 47 shows the resulting goal hierarchy. Each of the alternatives (STS) has to be connected with each of the free-nodes of the goal hierarchy at the left hand side.

We assume that T. Saaty's AHP is well known [52]. We show two examples of the necessary pairwise comparison matrices which have been elaborated by experts using Saaty's 9-point scale. Please consider that these matrices include data which are more artificially than real.

1) Pairwise comparisons of the first-level goals with respect to the top-goal: optimal STS

The matrix is reciprocical but not completely consistent. With $n = 8$ (number of goals) for the consistency-value KW yields: $0.097 < 0.1$, this means that we have sufficient

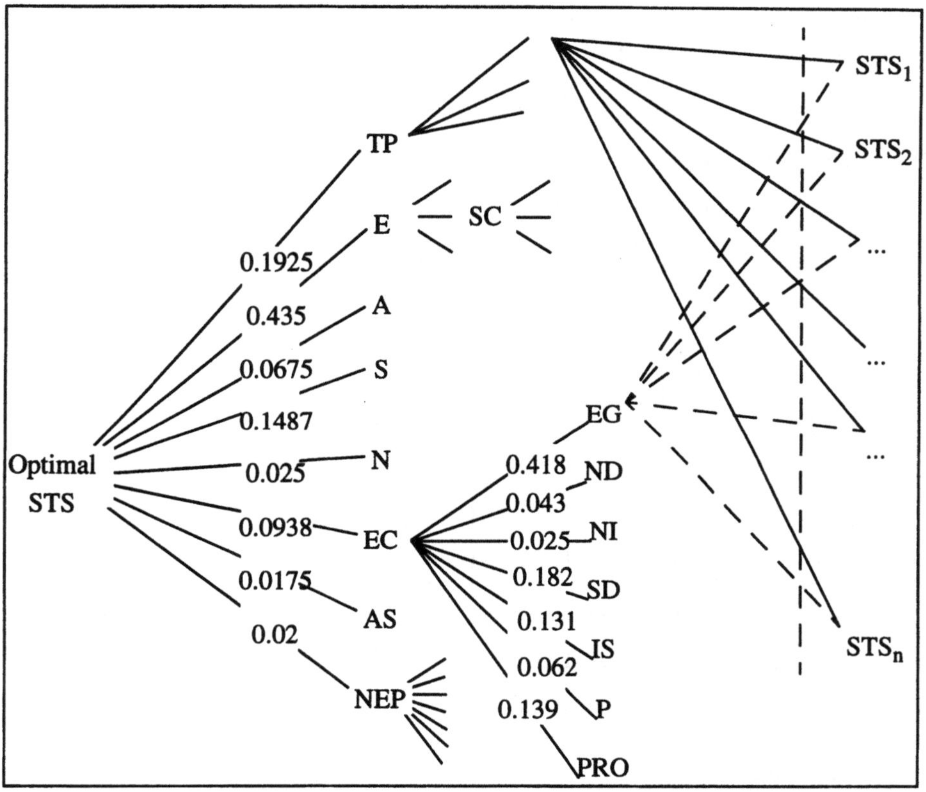

Figure 47: Part of the goal hierarchy and alternatives (STS's)

Top Goal	TP	E	EC	A	S	N	AS	NEP
TP	1	1/5	5	5	1	9	7	8
E	5	1	7	8	5	9	9	9
EC	1/5	1/7	1	2	1/2	7	7	7
A	1/5	1/8	1/2	1	1/3	8	4	5
S	1	1/5	2	3	1	7	9	9
N	1/9	1/9	1/7	1/8	1/7	1	1	1
AS	1/7	1/9	1/7	1/4	1/9	1	1	3
NEP	1/8	1/9	1/7	1/5	1/9	1	1/3	1

consistency. Then, the normalized vector w of weights of the first goal level is computed by the eigenvalue method

$$w = (0.1925, 0.435, 0.0938, 0.0675, 0.1487, 0.025, 0.0175, 0.02)$$

This might be assigned to the respective arcs in Fig. 47.

2) We consider the subgoals of EC and compare them pairwise with respect to EC.

EC	EG	ND	NI	SD	IS	P	PRO
EG	1	6	7	4	5	5	5
ND	1/6	1	4	1/5	1/4	1/3	1/4
NI	1/7	1/4	1	1/5	1/6	1/4	1/6
SD	1/4	5	5	1	2	4	2
IS	1/5	4	6	1/2	1	4	1
P	1/5	3	4	1/4	1/4	1	1/5
PRO	1/5	4	6	1/2	1	5	1

With $n = 7$ the consistency value of this matrix is $KW = 0.095 < 0.1$ (sufficiently consistent). The vector of normed weights of the subgoals of EC with respect to Environmental Computability is:

$$w = (0.418, 0.043, 0.025, 0.182, 0.131, 0.062, 0.139)$$

There are some disadvantages when using the crisp AHP.

- Some of the criteria are not crisp by their nature or could be modelled to be imprecise. For example Safety, Availability, Necessity, National/European Prestige should be modelled by fuzzy techniques rather than to try to find crisp subgoals (criteria) etc. Noise duration is a typical example, which one could model to be crisp. But this precision is not needed at all. Therefore a linguistic variable approach would meet better the granularity of a concept like noise intensity (Fig. 48).
- The use of Saaty's 9-point scale to determine the pairwise comparisons seems to be imprecise because of the underlying meaning of the numbers used. But this imprecision is not modelled, and the results do not take into account the imprecision of the pairwise comparisons. Therefore Fuzzy-AHP, using linguistic ranking, etc. [5] seem to be the more appropriate (but also more complicated,

memory and time consuming) approaches. Results in this direction will be published in the future.

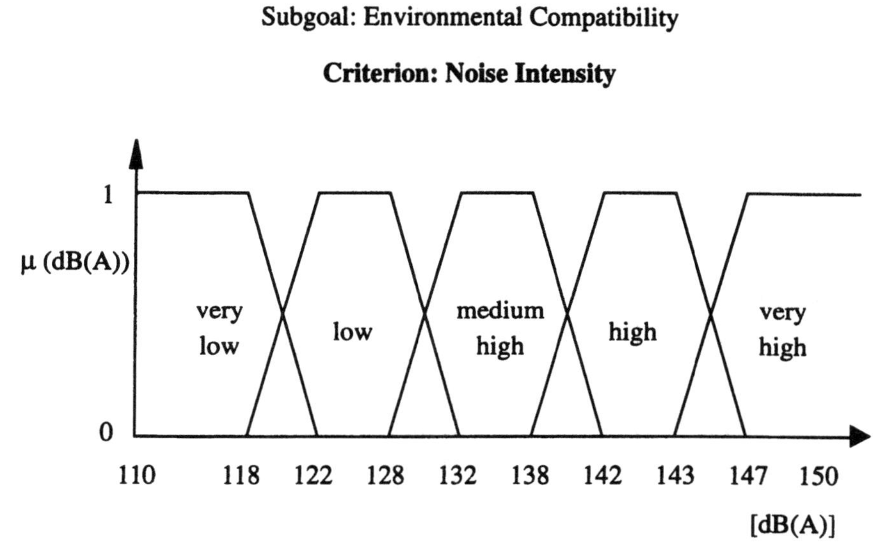

Figure 48: Linguistic Variable "Noise Intensity"

5.3. Logistic Layout

In this paragraph we will show how the specialized logistic system "Transportation of Letters" from Section 3.3 might be considered as a "layout-design problem". In contrast to the aircraft cabin layout problem the layout decisions are:

- choose hubs from a predefined set,
- choose vehicles and round-trips of the vehicles.

Using again KONWERK to configure the layout we have the domain-object hierarchy defined in Fig. 49.

Now, similar to the examples in Sections 5.1 and 5.2 one has to define the domain-objects and the conceptual constraints in order to create the domain-knowledge-base.

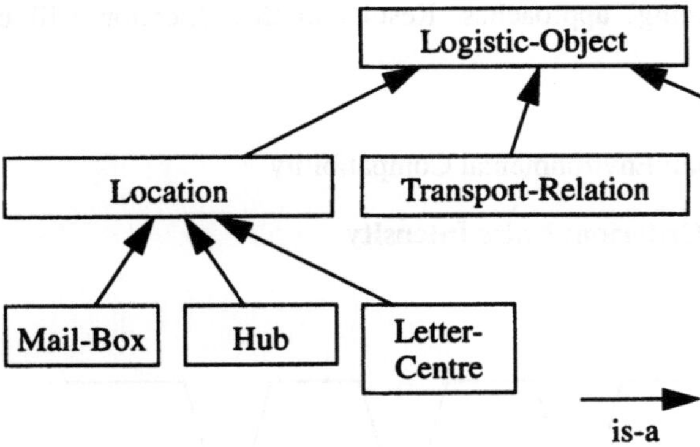

Figure 49: Specialization Hierarchy of Logistic Objects

Examples:

(def-do *:conceptname* mailbox
:superconcept location
:parameter ((expected-content [0 100])
(stoptime [0 100])
(number-in-tour [INTEGER 0 100]))
:relations (*is-location-of* (*a* hub)))

Therefore, the parameters characterizing a mailbox are the expected content (e.g. 60 letters), the time a car has to stop at the mailbox to do the necessary service and a number the mailbox will get assigned in the tour containing this location. The relation means that a mailbox has to be assigned to a hub. From the user-interface one is able to input specific data of any of the objects e.g. the mailboxes. Then, from those data and the declareated objects in the knowledge base one might get individuals by specializations like, e.g.:

(def-do *:conceptname* mailbox-mailstreet
:superconcept mailbox
:parameter ((expected-content 60)
(stoptime 5)
(number-in-tour [INTEGER 0 100]))
:relations (*is-location-of* #(HUB)))

The part-of-hierarchy (compositional hierarchy) assigns objects to more complex objects. Fig. 50 shows a simplified compositional hierarchy using the notation from Section 2.1.

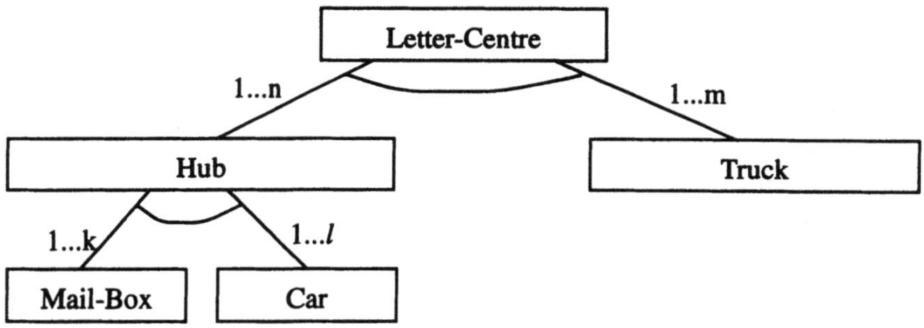

Figure 50: Simplified Compositional Hierarchy

The computation of costs is realized mainly by constraints, which are automatically transformed to the optimization-model base. The following is an example of a conceptual constraint which computes the overall-cost of a hub by summation of costs for the transportation-to and transportation-back from the hub and the costs for building and using the hub.

```
(def-conceptual-constraint
        :name Overall-Cost-Hub
        :variable-pattern-pairs
                        (?H: name hub)
        :formula        ("?H.Sum-Cost =
                        ?H.Cost-to + ?H.Cost-back +
                        ?H.Cost-build * ?H.Existence"))
```

Now we briefly describe how the logistic system is designed using this domain knowledge-base. The solution will be generated by a 3-phase approach:

Phase 1: Computation of the best transport-relations beween locations using Dijkstra's algorithm. This computation is done within the application graphical user-interface.

Phase 2: Clustering of the mail-boxes to groups (classes) and assigning each group to a hub.

Phase 3: Within each of the groups vehicle routing problems are solved by several heuristics, e.g. by the savings heuristic. Doing this the number of vehicles which are necessary to fulfill the transportation tasks will be computed.

The reason a 3-phase overall heuristic approach is choosen is the complexity of the problem. Of course a combinatorial optimization-model which describes the overall problem is possible to elaborate, but it is impossible to solve it even for small-size applications.

Phase 2:
For the selection of the hubs and the assignment of mailboxes to hubs (location problem; it is not necessary to use each of the potential hubs) we use and solve an optimization problem. The mailboxes we assigned to hubs by minimization the costs of to-and-back tours only. This is of course an approximation because in reality round-trip-tours are used. The quality of the approximation increases if there are time windows and mass-constraints for the transportation. Also the quality is much better than a heuristic clustering on the basis of Eucleadean distances, because the underlying street network is used (and not Eucleadean distances) and time- and mass-constraints are also considered.

The resulting MIP (Mixed Integer Problem) is solved by the MOMIP [18] software system. The results of the optimization are given back to the object-slots in the knowledge-base. For example any box has, after the Phase 2 optimization, the name of a hub in the relations-slot. That makes it possible to generate all compositional relations between hubs and mailboxes.

Another alternative would be to use fuzzy clustering in the second phase instead of the method described.

Phase 3
In this phase, for each of the hubs, there are vehicle-routing problems defined in the knowledge base. Each of them is characterized by a hub, the assigned mailboxes, the transport relations in the street network, an objective function (cost-function) and time- and mass-constraints. These problems are solved by vehicle routing heuristics [27]. Currently the Savings heuristic [18] is available only in KONWERK. The system described above was implemented by M. Thärigen. Fig. 33 shows a hard copy of the graphical interface which visualized a solution.

A next step to extend the model is to incorporate fuzzy modelling. There are several alternative methods to use fuzzy logic (in particular for the Phase 2 of the approach):

- A classification of the mailboxes using linguistic variables to measure the "expected" number of letters in the mailbox: e.g., small, medium, big mailboxes. An alternative is to represent the "expected number of letters" in a mail-box by fuzzy numbers.
- The transportation-time from one location to another one could be also modelled by fuzzy numbers to take into account the uncertainty of those data.

- Finally goals could be transformed into fuzzy goals.

It is not always sensible to make the decision on the basis of exact costs. Therefore, one could introduce (context dependent) linguistic variables for costs and to define fuzzy goals on that basis.

The flexibility of services of mailboxes would be another example for a fuzzy (possibility) goal. It has to be strongly forbidden that a letter collecting car visits a mailbox before the service interval has started. However, using a fuzzy set "prefered-service-time" one could model preferences for each of the mailboxes (Fig. 51).

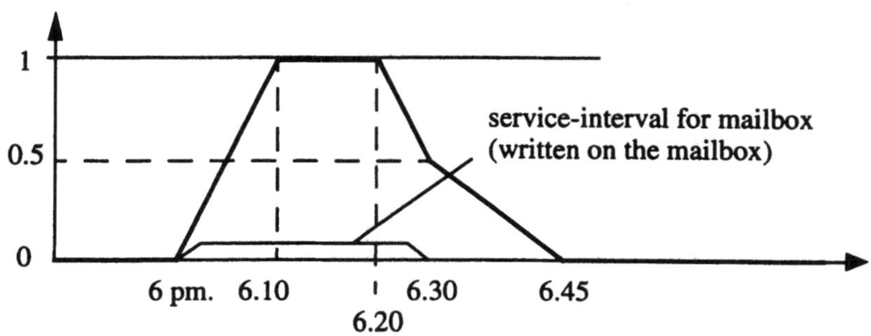

Figure 51: Membership function of a fuzzy set "prefered-service-time" (mailbox)

5.4. Configuration of a PC by an Evolutionary Algorithm - The Fuzzy-Multi-Criteria Fitness Evaluation

The methodology described in the following was proposed and implemented by T. Kriese [25]. Repeating from Section 3.4 we assume that the genetic kernel of the algorithm generates n individuals (consistent PC's) per generation. Each individual has m basic-attributes. These are the attributes of the PC components, e.g., prices and technical type data. The task of the evaluation module is to evaluate the fitness of each of the individuals using the user requiments. Any evaluation is ultimately based on the basic-attributes.

As user requirements we consider, similar to the requirement modelling in Section 3 three types:

- global optimization criteria,
- structural-type user requiremements,
- reservation levels.

Global Optimization Criteria

The criteria (also objectives, goals) are referred to as global because they are more general than focused on PC-configurations. These criteria are

1. Price
2. Fexibility/Extendability
3. Actuality
4. Professionality
5. Overall-Performance
6. Quality.

The user might give an indirect weighting of those criteria by pairwise comparison, while the vector of weights will be computed by Saaty's eigenvalue-method [52].

Structural-type User Requirements

A PC has, at a minimum, to contain all necessary components to enable it to work. Optional components extend the minimal-configuration defined by the necessary components. Optional elements could be a Mouse or a CD-ROM. If the user requires optional components, then those components will be redefined from the evolutionary algorithm to become necessary-type components. As a result, each consistent configuration will have those additional components and therefore structural requirements need not to be considered within the evaluation module. The only problem which might occur is that the user prescribes an incomplete model such that the system is unable to generate a feasible configuration. However, this danger might easily be omitted by checks in the user interface.

Reservation Levels

Reservation levels express "minimal-requirements" for the respective attributes and might be formulated on both the component and the building-block level. Specializations which don't fulfill the reservation level requirements will be considered to be inconsistent.

1) Component level

Each component has its basic-attributes. In many cases, the basic-attributes have discrete value-sets.

- For each basic-attribute the user can formulate a reservation level (a minimal-requirement which should be at least reached).
- The difference between the value of the attribute and the reservation level will be used for the evaluation of the component. If there is no reservation level specified for a determined basic-attribute, then the difference will be specified to be zero.

As an example of a reservation-level we consider: The Graphic Card should have the maximal resolution of at least 1024 x 768 pixels.

2) Building-block Level
At the building block level the attributes are mostly composed of basic-attributes. Examples are Graphic-System-Performance, Data-Export-Abilities, Multi-Media-Abilities (Audio, Video etc.).

Because the user would have difficulty formulating a reservation level for the Graphic-System-Performance between 0 and 1, we define reservation levels at the building-block level by linguistic values. Using linguistic variables we have a comfortable tool to define a set of values (terms) which might be attended by the reservation levels. For example the user might require the reservation level "very high" for the Graphic-System-Performance.

Similar to the component level there is a "difference" between the reservation level and the real value of the respective attribute of the considered building block. These "differences" are used for the evaluation. The computation of "appropriate differences" is based on Chen and Hwangs method [5]. Using the 8-scales of Chen and Hwang to represent the linguistic values and the transformation of the linguistic into real values we are able to compute the desired differences.

To briefly illustrate the approach described above we discuss an example below:

Example: Graphic-System Performance
We consider the composed attribute "Graphic-System Performance". The linguistic variable to describe reservation levels for this attribute has the terms:

{very-good, good, medium, bad, very-bad}

The following components and their attributes influence the Graphic-System Performance:

Component	Graphic Card	Monitor
basic-attributes	maximal resolution number-of_colours memory bus system	monitor-diagonal vertical-frequency maximal-resolution

For each of the linguistic terms of the composed attribute and for each of the basic attributes we define fuzzy sets by setting the respective membership function values. In

Table 6 we illustrate such a definition for the four attributes of "Graphic Card". The membership-values of *very-good, bad* and *very-bad* are computed (as usual) by fuzzy-modifiers [54].

Table 6: Maximal resolution of Graphic Card

maximal-resoluion	640 x 480	800 x 600	1024 x 768	1280 x 768
$\mu_{very\text{-}good} = (\mu_{good})^2$	0.04	0.36	1.0	0.81
μ_{good}	0.2	0.6	1.0	0.9
μ_{medium}	0.7	1.0	0.7	0.3
$\mu_{bad} = 1 - \mu_{good}$	0.8	0.4	0.0	0.1
$\mu_{very\text{-}bad} = (\mu_{bad})^2$	0.64	0.16	0.0	0.01

Table 7: Number of Colours

number-of-colours	2	16	256	65000	Truecolour
$\mu_{very\text{-}good} = (\mu_{good})^2$	0.0	0.04	0.36	0.81	1.0
μ_{good}	0.0	0.2	0.6	0.9	1.0
μ_{medium}	0.0	0.5	1.0	0.6	0.0
$\mu_{bad} = 1 - \mu_{good}$	1.0	0.8	0.4	0.1	0.0
$\mu_{very\text{-}bad} = (\mu_{bad})^2$	1.0	0.64	0.16	0.01	0.0

Table 8: Working Memory

working-memory	0.512	1	2
$\mu_{very\text{-}good} = (\mu_{good})^2$	0.25	0.64	1.0
μ_{good}	0.5	0.8	1.0
μ_{medium}	0.7	1.0	0.4
$\mu_{bad} = 1 - \mu_{good}$	0.5	0.2	0.0
$\mu_{very\text{-}bad} = (\mu_{bad})^2$	0.25	0.04	0.0

Using these fuzzy sets of the basic-attributes of Graphic Card we now define the respective fuzzy sets of the Graphic Card with respect to the Graphic-Performance. The worst attribute of a Graphic Card is always the bottleneck with respect to the Graphics

Table 9: Bus System

bus system	ISA	VL	PCI
$\mu_{very\text{-}good} = (\mu_{good})^2$	0.25	0.64	1.0
μ_{good}	0.5	0.8	1.0
μ_{medium}	0.9	0.2	0.1
$\mu_{bad} = 1 - \mu_{good}$	0.5	0.2	0.0
$\mu_{very\text{-}bad} = (\mu_{bad})^2$	0.25	0.04	0.0

performance of the Card. Therefore to aggregate the four attributes we use the Min-operator:

$$\mu_i(GC) = Min(\mu_i(GC.resolution), Min(\mu_i(GC.colours), Min(\mu_i(GC.working\text{-}memory), Min(\mu_i(GC.bussystem)))))$$

(GC means any of the Graphic Cards, $i \in \{$very-good, good, medium$\}$

On the other side the fuzzy sets representing bad or very-bad performance of the Graphic Card have to be computed by using the Max-operator from the components.

$$\mu_i(GC) = Max(\mu_i(GC.resolution), Max(\mu_i(GC.colours), Max(\mu_i(GC.working\text{-}memory), Max(\mu_i(GC.bussystem)))))$$

(GC means any of the feasible Graphic Cards, $i \in \{$bad, very-bad$\}$

In Table 10 we define the membership-functions with respect to the basic-attributes of a Monitor.

Table 10: Monitor-Diagonal

monitor-diagonal	14	15	17	19	21
$\mu_{very\text{-}good} = (\mu_{good})^2$	0.36	0.49	1.0	0.81	0.64
μ_{good}	0.6	0.7	1.0	0.9	0.8
μ_{medium}	0.8	1.0	0.3	0.8	0.6
$\mu_{bad} = 1 - \mu_{good}$	0.4	0.3	0.0	0.1	0.2
$\mu_{very\text{-}bad} = (\mu_{bad})^2$	0.16	0.09	0.0	0.01	0.04

Using the same reasoning we made considering the Graphic Card we get for any Monitor MO the membership functions by aggregation using the Min- or the Max-operator respectively.

Table 11: Vertical Frequency

vertical-frequency	50	70	100
$\mu_{very\text{-}good} = (\mu_{good})^2$	0.0	0.81	1.0
μ_{good}	0.0	0.9	1.0
μ_{medium}	0.5	0.2	0.0
$\mu_{bad} = 1 - \mu_{good}$	1.0	0.1	0.0
$\mu_{very\text{-}bad} = (\mu_{bad})^2$	1.0	0.01	0.0

Table 12: Maximal resolution

maximal-resoluion	640 x 480	800 x 600	1024 x 768	1280 x 768
$\mu_{very\text{-}good} = (\mu_{good})^2$	0.04	0.36	1.0	0.81
μ_{good}	0.2	0.6	1.0	0.9
μ_{medium}	0.7	1.0	0.7	0.3
$\mu_{bad} = 1 - \mu_{good}$	0.8	0.4	0.0	0.1
$\mu_{very\text{-}bad} = (\mu_{bad})^2$	0.64	0.16	0.0	0.01

$$\mu_i(MO) = Min(\mu_i(MO.diagonal), Min(\mu_i(MO.vertical\text{-}frequency), Min(\mu_i(MO.maximal\text{-}resoluion))))$$

$$MO \in Monitor\text{-}Set,\ i \in \{very\text{-}good, good, medium\}$$

$$\mu_i(MO) = Max(\mu_i(MO.diagonal), Max(\mu_i(MO.vertical\text{-}frequency), Max(\mu_i(MO.maximal\text{-}resoluion))))$$

$$MO \in Monitor\text{-}Set,\ i \in \{bad, very\text{-}bad\}$$

Now, we have fuzzy sets for very-good, good etc. Monitors or Graphic Cards with respect to their graphics performance. Now, we aggregate these fuzzy sets in order to get an overall Graphic-System Performance of the PC (better, of the building block Graphic Card-Monitor). We are using the Gamma operator to do this aggregation in order to take into account the compensatory ability of this operator [54].

In our application compensation means recompense for good graphic quality. The parameter Gamma quantifies this recompense.

We get $\mu_i(GS) = (\text{Min}(\mu_i(GS.MO), \mu_i(GS.GC))^{1-\gamma}$
$* (\text{Max}(\mu_i(GS.MO), \mu_i(GS.GC))^{\gamma})$
where $0 \leq \gamma \leq 1$, $i \in$ {very-good, good, medium, bad, very-bad}
$GS.MO \in$ Monitor-Set, $GS.GC \in$ Graphic Card-Set

Now we consider a computational example:
There should be given a Graphic-System with the following technical data:

Monitor: resolution = 800 x 600, diagonal = 14"
vertical-frequency = 50 Hz

Graphic Card: resolution = 640 x 480, colours = 256
working-memory = 0.512 MB, bussystem = ISA

Then, looking to the tables, we get the following membership values:

- Monitor: $\mu_{very\text{-}good}(MO) = \text{Min}(\mu_{very\text{-}good}(14),$
$\text{Min}(\mu_{very\text{-}good}(50), \mu_{very\text{-}good}(800x600))) = 0.0$

and analogously

$\mu_{good}(MO) = 0.0$, $\mu_{medium}(MO) = 0.5$
$\mu_{bad}(MO) = 1.0$, $\mu_{very\text{-}bad}(MO) = 1.0$

- Graphic Card: $\mu_{very\text{-}good}(GC) = \text{Min}(\mu_{very\text{-}good}(640x480),$
$\text{Min}(\mu_{very\text{-}good}(256), \mu_{very\text{-}good}(ISA))) = 0.04$

and analogously

$\mu_{good}(GC) = 0.2$, $\mu_{medium}(GC) = 0.7$
$\mu_{bad}(GC) = 0.8$, $\mu_{very\text{-}bad}(GC) = 0.64$

With $\gamma = 0.5$ we get for the Graphic-System GS

$$\mu_{very\text{-}good}(GS) = (\text{Min}(0.0, 0.04))^{0.5} * (\text{Max}(0.0, 0.04)^{0.5} = 0.0$$

$$\mu_{good}(GS) = (\text{Min}(0.0, 0.2))^{0.5} * (\text{Max}(0.0, 0.2)^{0.5} = 0.0$$

$$\mu_{medium}(GS) = 0.5^{0.5} * 0.7^{0.5} = 0.591$$

$$\mu_{bad}(GS) = 0.8^{0.5} * 1.0^{0.5} = 0.894$$

$$\mu_{very\text{-}bad}(GS) = 0.64^{0.5} * 1.0^{0.5} = 0.8$$

Using a special defuzzification (maximal membership) we could now assign the linguistic value "bad" as Graphic-System performance of the given GS. Another possibility would be to consider the nearest neighbour to the maximal membership. This would result in

$$\mu_{bad} - \mu_{medium} \geq \mu_{bad} - \mu_{very\text{-}bad}$$

The conclusion would than be to evaluate the GS with "bad to very bad", which would mean to add linguistic intermediate values. If the user had defined an aspiration level "very good", then we would have to compute the difference using a transformation-scale of Chen and Hwang which consists in the linguistic values very-good and bad. This difference would then to be used as an attribute of the evaluation of the individual.

The fitness-vector FIT(POP) of a population might be defined using several different appoaches. Each component of $(FIT(POP))_i$ represents the fitness of the ith individual belonging to the population (n-individuals, constant).

The approach proposed by T. Kriese uses a vector w of weights and a matrix A resulting in

$$FIT(POP) = A \cdot w$$

The vector of weights is defined by

$$w^T = (w_1, ..., w_6 ; -bg_1, ..., -bg_n; -bt_1, ..., -bt_m)$$

where

* $w_1, ..., w_6$ are the weights of the global-evaluation criteria (computed by Saaty's eigenvalue method)
* $bg_1, ..., bg_n$ are the weights of the differences of the building-block attributes from the respective reservation levels.
* $bt_1, ..., bt_m$ are the weights of the differences of the components basic-attributes from the respective reservation levels.

(If there is no such aspiration level defined, then the respective weight is zero. The weights are not changeable by the user.)

The minus in the definition of w in relation with $A \cdot w$ means that for each individual the evaluation is a weighted sum of the global criteria minus a weighted sum of the differences in the reservation levels.

The rows of the matrix A contain the individuals (alternatives for selection). The columns contain all the attributes of the individuals. All matrix-elements are real and normed. Table 13 shows a listing of all attributes used in the experimental version of the system (global criteria and differences from the reservation levels).

For each of these attributes a reservation level might be defined. The difference in the respective value of the individual considered is then used in the evaluation matrix A. A discussion of all the basic attributes of components will be not presented here in order to keep the example small enough.

At the building-block level the following composed attributes are used:

Table 13: Composed Attributes

building block notation	component notation	necessary/optional
Memory group	FDD FDD HDD HDD Tapedrive Controller	n o n o o n
Graphic system	Monitor Graphic card	n n
Multimedia-group	Soundcard Videocard CD-ROM Video-accelerator-kit Video-player Loudspeaker	o o o o o o
Computing-group	Motherboard Multi-IO-card RAM Cache Coprocessor Power supply Enclosure Processor	n o n o o n n n
Input-group	Keyboard Mouse Joystick Tablet Scanner Barcode-scanner	n o o o o o
Output-Group	Printer	o
Telephone-connection	Modem	o
Network-group	Network-card	o

5.5. Model-Configuration Using an Example from the Environmental Research Area

The first aim of this subchapter is to show "how to model requirements". We use the RAINS model to illustrate this approach. A second aim is to briefly introduce an "overall modelling approach" and to illustrate how a modelling support system should be designed and how it should work.

5.5.1 The Requirement Model including Imprecision

In technical or engineering type domains there are three main types of user requirements:

(1) functional requirements,

(2) global constraints (formulated by the user),

(3) objectives (also called goals, aims etc.).

In [57] we introduce requirement modelling including imprecise requirements for engineering type of applications. For modelling purposes in the environmental research area the second and third type (2), (3) of the user-requirements seem to be adequate, however functional type requirements (1) could also be introduced.

In contrast to the conceptual constraints which describe the logical relations and natural restrictions, we formulate in the following the

requirement-type-constraints,

which are user-formulated constraints that have to be added to the core model in order to restrict the solution space.

Typically, these constraints (and later also the goals) should be formulated using the attributes introduced in the conceptual hierarchy. In this sense, the conceptual hierarchy serves as a source of the requirement modelling.

In the following we will illustrate the approach first by considering the RAINS-knowledge model, briefly introduced in Section 2.

Using the information given in Fig. 52 and 53 we can select the following attributes given by slot-names and filler-symbols:

emission-level-of-source-region k:x^k

emission-level-vector:x

cost-of-reducing-emission-to-level x^k:$f_k(x^k)$

cost-of-overall-concept:C

grid-deposition-of-grid j:y_j

grid-deposition-vector:y

(For more flexibility we consider attributes of concepts rather than of individuals for example EUROPE and it's parts.)

Now, having those attributes, the user might wish to define bounds such as:

maximal-grid-deposition-of-grid j:b_j

minimal-emission-level-of-source-region k:l_k

maximal-emission-level-of-source-region k:u_k

maximal-cost-of-reducing-emission-to-level x^k:Cu_k

Of course, such bounds lead to user formulated constraints that are caused by scientific or political requirements.

These are examples of user defined constraints, which added to the core model restrict the solution space and define a "requirement model". Before we show in more detail how to get this "requirement model" using an interactive approach, we will touch briefly on the aspect of imprecision. To start introducing the quantitative concepts of imprecision, which are well known in the area of Fuzzy Logic, we consider the bound-constraints (*).

$$(*) \quad \begin{cases} y_j \leq b_j & \text{for all or for a subset of grids } j \\ l_j \leq x^k \leq u_k & \\ f_k(x^k) \leq Cu_k & \end{cases} \Big\} \text{ for all or for a subset of source regions } k$$

Assume, the cost-function $f_k(x^k)$ to be given for a certain source-region k (Fig. 52).

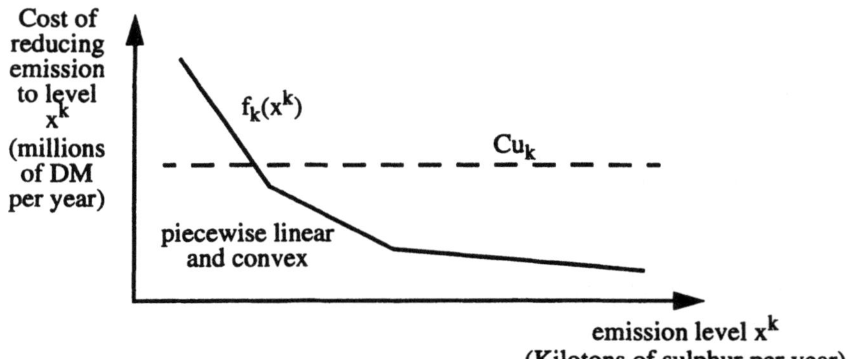

Figure 52: Cost-function $f_k(x^k)$ for a source-region k

Naturally, there is a lot of uncertainty in the cost-function. A method for dealing with this uncertainty using probabilities is presented in [3]. However, even if $f_k(x^k)$ is the real cost-function, or if we would know the probabilities of the finite number of scenarios proposed in the stochastic approach [3], no one knows exactly how to set-up all the bounds Cu_k, l_k, u_k, b_j.

Alternatively, by relaxation of the crisp constraints (*) and using fuzzy constraints, we are able to help the user to define softer bounds. We illustrate this approach by considering the crisp constraint $y_j \leq b_j$.

We define the fuzzy set:

$$\text{A-G-D}_j = \text{"Acceptable-Grid-Deposition-at-Grid-j"} = \{(y_j, \mu_j(y_j)) \mid y_j \in R^+\}$$

where $R^+ = \{y_j \in R \mid y_j \geq 0\}$ is a crisp basis set and

$$\mu_j(y_j) = \begin{cases} 1, & \text{if} \quad y_j \leq b_j \\ \in [0, 1], & \text{if} \quad b_j \leq y_j \leq b_j + \varepsilon_j \\ 0, & \text{otherwise} \end{cases}$$

$$\varepsilon_j > 0$$

describes a membership-function μ_j over R^+ (Fig. 53).

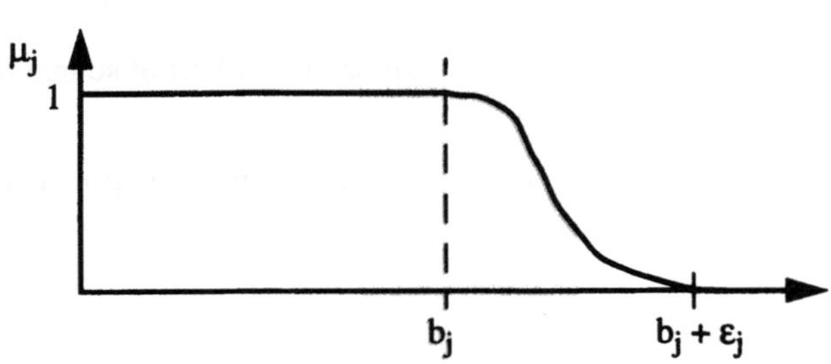

Figure 53: Membership function for "Acceptable-Grid-Deposition"

Using the fuzzy constraint

"$y_j \in \text{A-G-D}_j$", shortly denoted by $y_j \lesssim b_j$

b_j becomes the meaning of maximal deposition y_j which is completely acceptable ($\mu_j = 1$). In the interval $[b_j, b_j + \varepsilon_j]$ we have a smooth transition from completely acceptable to completely unacceptable (infeasible, $\mu_j = 0$) deposition, which is attained first for $y_j = b_j + \varepsilon_j$.

In the same way we can deal with the cost $f_k(x^k)$ to introduce a fuzzy set

$$\text{A-C-RE}(x^k) = \text{"Acceptable-Cost-of-Reducing-Emission-to-level-}x^k\text{"}$$
$$= \{(f_k(x^k), m_{AC_k}(f_k(x^k))) \mid f_k(x^k) \in R^+\}$$

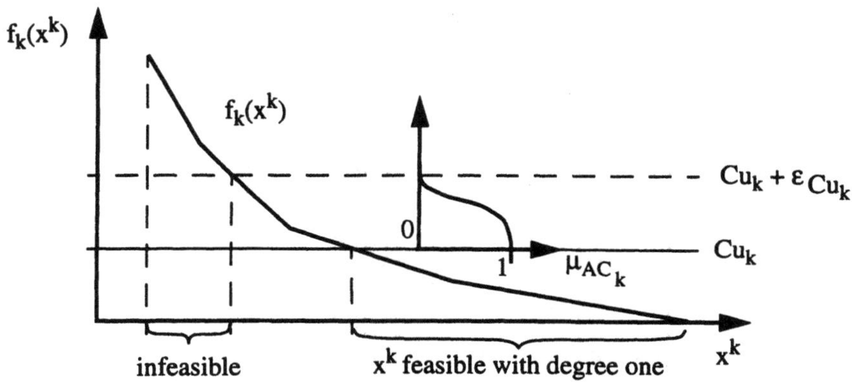

Figure 54: Membership function for "Acceptable-Cost"

Another way to model imprecise user requirements quantitatively is to define linguistic variables.

We have introduced two examples of fuzzy sets above. Using linguistic variables we have to formulate several fuzzy sets in addition/in order to model the users imprecise linguistic terms. For example:

Emission-level-of-Source-Region-k
$\in$ {very-low, low, medium, high, very-high}

Deposition-at-Grid-j $\in$ {small, medium, high, unacceptable-high}

Cost-of-Reducing-Emission-to-level-x^k
$\in$ {very-low, low, not-high, moderate, not-low, high, very-high}

The definition of the "set of terms" has to be done carefully with respect to the linguistic variable under consideration. In addition, the quantification of such linguistic terms as low, high, medium etc. using fuzzy sets is context-dependent (Fig. 55).

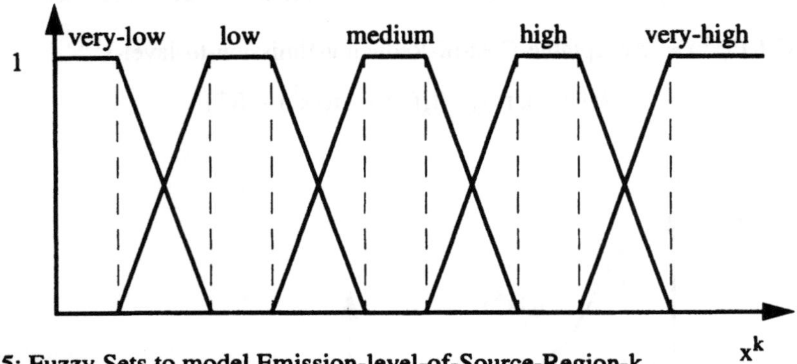

Figure 55: Fuzzy Sets to model Emission-level-of-Source-Region-k

In a similar way, we can define the fuzzy-sets related to the linguistic variables:

Deposition-at-Grid-j

Cost-of-Reducing-Emission-to-level-x^ketc.

If there are real measurements of emissions x^k, depositions y_j and costs $f_k(x^k)$ one can use the linguistic variables to build an user interface dealing with such imprecise terms. If, for example, the deposition at grid j is known to be 0.85 (g of sulphur pro m^2 and year), then the membership degree to the fuzzy set "Small-Deposition" is 0.5. Also, the membership degree to the fuzzy set "Medium-Deposition" is 0.5 provided the fuzzy sets are defined as shown in Fig. 56.

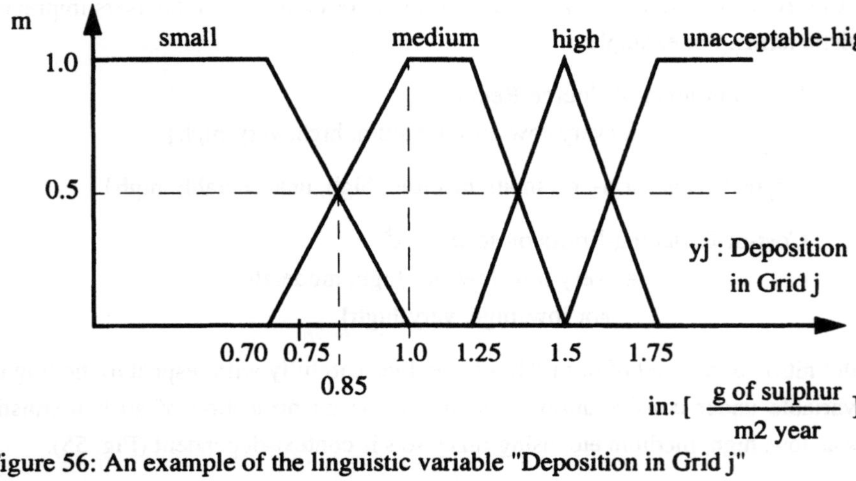

Figure 56: An example of the linguistic variable "Deposition in Grid j"

If there is a common acceptance of a definition similar to that shown in Fig. 56, an inexperienced user need not know the definition of the fuzzy sets to start to use such a model. He can formulate requirements like "the deposition should be small" or "the deposition should be not high or larger than high".

Then, the system translates these requirements into fuzzy-constraints using the fuzzy sets and operations with fuzzy sets [54]. Of course, the user might consider the resulting fuzzy constraint and decide if it really models the requirement he had in mind. Then, he might start to change the requirement iteratively if necessary.

Similarly, one might use the linguistic variables to deal with "possibilities" [54]. Let us assume that there are no measurements of the deposition at grid j available. But there are expert estimations like "the deposition is small". Then, the membership function of the fuzzy set "small-deposition" defines the possibility of a value, assuming "small deposition" is known to be true.

For example assuming "small deposition" $y_j = 0.5$ or $y_j = 0.7$ have the possibility value 1.0. The deposition $y_j = 0.85$ has the possibility of 0.5 and $y_j = 1.0$ the possibility 0 (Fig. 56).

In the following we show how linguistic variables might be used to construct fuzzy goals or fuzzy constraints.

(1) The use of fuzzy goals

To introduce fuzzy goals, we start by considering a crisp objective, as for example:

$$C(x) = \sum_k f_k(x^k) \quad \rightarrow \quad \text{Min.} \qquad (**)$$

The idea in the crisp case is to choose x^k, $k = 1(1)k$ such that C(S) would be minimized provided x^k fulfills the set of all constraints.

If constraints are replaced by fuzzy constraints it makes no sense to continue to utilize a crisp objective, because it will be difficult to compare solutions x. Some might be completely feasible $\mu = 1.0$, but the value C(x) might be bigger (meaning worse) than the value $C(x^*)$ of a solution that is feasible with $\mu = 0.8$ only.

Therefore, the classical Bellman-Zadeh approach suggests we replace (**) by $C(x) \leq C_u^*$ and consider this inequality as a fuzzy constraint.

There are several methods to construct the fuzzy set in Fig. 57 for the cases with only one or with multiple goals [52], [53], [55].

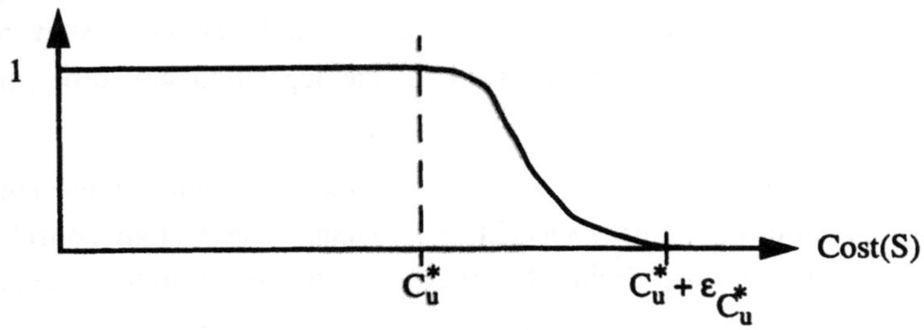

Figure 57: Membership function to model an objective as fuzzy constraint

.Another way to introduce fuzzy goals is to consider linguistic variables. To illustrate this approach, let us assume that there is for each k a linguistic variable:

"Cost-of-Reducing-Emission-to-Level-x^k"

with the definition given in Fig. 58.

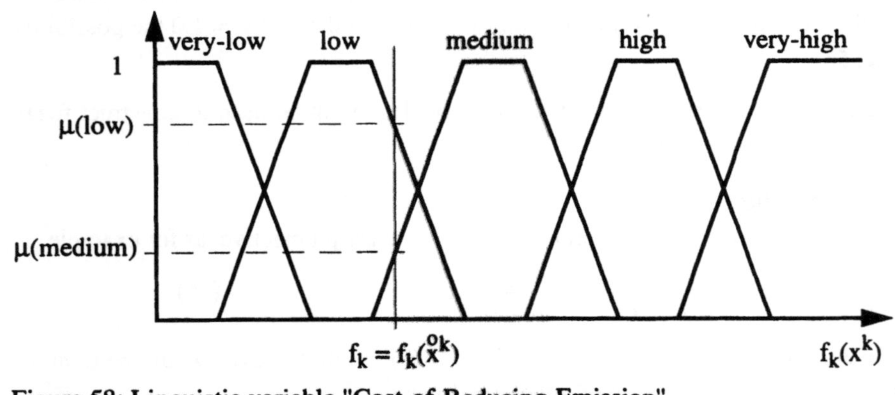

Figure 58: Linguistic variable "Cost-of-Reducing-Emission"

Of course, this definition depends on the region k under consideration (context dependent). x^k does not need to be known to define these fuzzy sets, but one should know the range of possible values of f_k for each k and the meaning of low, high etc. These fuzzy sets have to be added to the respective slot-description of the concepts "Source-Region-k" (Section 4 illustrates how it has been implemented in KONWERK). If $\overset{o}{x}{}^k$ is known, then because of the crisp function $f_k = f_k(\overset{o}{x}{}^k)$ there are known membership-values of the respective linguistic terms.

Now, considering the linguistic variables for each k we are able to define a fuzzy set

$$F_k(t_1(k), \ldots, t_{N_k}(k))$$

where $t_1(k), \ldots, t_{N_k}(k)$ denote the fuzzy sets assigned to the terms of the linguistic variable

"Cost-of-Reducing-Emission-to-Level-x^k".

For example, for Source Region k = 1, one could choose:

F_1 = (low-cost) or (medium-cost),

but for k = 2 the function could be:

F_2 = ((medium-cost) or (high-cost)) and (not-very-high-cost)

The linguistic "or" or "and" can be interpreted by any of the operators known in the fuzzy logic [53], [55].

Now, as a fuzzy goal, we can introduce

$$F^* = \underset{k}{\text{AND}}\ F_k(t_1(k), \ldots, t_{N_k}(k))$$

The AND-operator should be interpreted to be L. Zadehs Min-Operator in order to forbid compensation between countries (source regions).

Unless we have to define optimality in such a case more precisely, we want to remark that an "optimal solution according to the fuzzy goal F*" could be defined to be a crisp vector $\hat{x}$ which maximizes the membership function of F*. However, there are also other possibilities.

<u>(2) Fuzzy IF-THEN-Rules</u>

The assumption we made to model "cost" by imprecise linguistic terms to leave the function $C_k = f_k(x^k)$ crisp and assumed to be well known, is of course not the only way to deal with cost. To model both the uncertainty and the imprecision more adequately one could consider the relation between x^k and C_k as a fuzzy relation or a fuzzy graph.

Also one could use the fuzzy-if-then rules approach to model the fuzzy relation. This would give us many possibilities to apply fuzzy inference and neuro-fuzzy-systems to learn (adapt) membership functions. To describe such approaches another paper is needed. However, to give a little insight we formulate examples of such fuzzy-if-then-rules which represent another type of relation between x^k and cost C_k.

IF (Emission-level-x^k-is-very-high),THEN (COST-C_k-IS-VERY-LOW)

IF (Emission-level-x^k-is-high),THEN (COST-C_k-IS-LOW <u>or</u>
COST-C_k-IS-VERY-LOW)

etc.

This sets of Fuzzy-IF-THEN-RULES might replace the crisp functions $C_k = f_k(x^k)$, in order to model the uncertainty and imprecision. Then, different methods of fuzzy inference (e.g. Max-Min-, Max-Product Inference) might be applied to conclude from imprecise knowledge on the emission level to (also imprecise) knowledge about the cost of reducing this emission. The fuzzy sets used to formulate prerequisites and conclusions of the rules and to some extend also the rules themselves are country-dependent. One method to find the rules could be by learning using examples.

5.5.2. The Overall Modelling Approach

Up to now, we have described several concepts to model the real existing imprecision in the models of the environmental research area illustrated using the RAINS model. However, we have only considered components of an optimization model (including imprecision).

Now we propose a general approach which describes how to get a model including MCDM and Fuzzy Logic by using the knowledge model we introduced in 2. (The Domain Model).

This approach called "the overall modelling approach" describes the "Modelling Methodology". Fig. 59 gives an overview, but in the following the main blocks are explained in more detail and subsequently two examples are considered to illustrate how the approach works.

<u>Comments on block [2]:</u>

- A finite set of explicitly given decision alternatives might be chosen by selecting a set of concepts (e.g. generic objects, individuals) from the Domain Knowledge Model [1] and checking the conceptual constraints to guarantee feasibility.
- The decision variables should be derived from the slot descriptions of the concepts belonging to both the conceptual and compositional hierarchy.

<u>Comments on block [3]:</u>

- The "Core Model" consists of <u>either</u>
  - the finite subset of feasible decision alternatives, where infeasibility means that all explicitly given alternatives, inconsistent with the conceptual constraints are omitted,

  or

  - a finite set of constraints formulated using the decision variables. These constraints are analytic-type reformulations (representations) of the conceptual constraints.

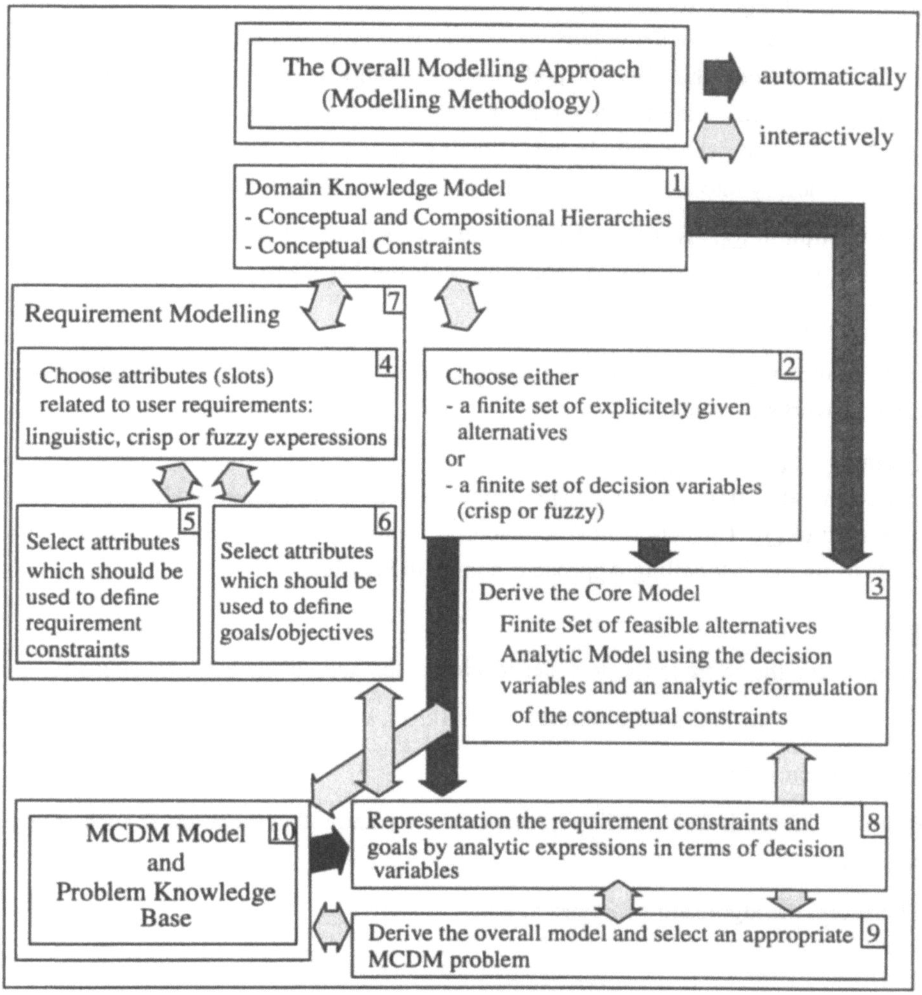

Figure 59: Architecture of the "Overall Modelling Approach"

- The "analytic-type formulation" of the constraints is widely used in the optimization and in the Operations Research field. In contrast to this type of representation the conceptual constraints in the knowledge base are modeled using a prefix notation and implemented quite often using LISP or C++. In KONWERK [8], [9] we generate the analytic-type representation automatically from the knowledge base and the decision variables.

Comments on block 8, 9, 10:

- Without describing the details we assume that there is a knowledge base which defines exactly the MCDM-model and it's components. It also contains a description of possible MCDM-problems related to the MCDM model. We will illustrate the approach briefly in the following by examples, however, without going more deeply into the details of the reasoning algorithms. To get the desired representation in 8, algorithms similar to the ones to get 3 from 1 and 2 have to be applied. In particular, a matching with the MCDM-model knowledge base 10 is necessary for both 3 and 8.

In the following, we give two examples to show how the "Overall Modelling Approach" works in principle.

Example 1 (The complete crisp case)
We assume the "domain knowledge model" to be given as introduced in Section 2 represented in 1. The task is now to elaborate a model for any arbitrary "overall concept" using the "Modelling Approach" shown in Fig. 59.

Block 2
Considering the "Source Region" description in the knowledge base, we choose

level-of-emission-of-source-region-k: x^k, $x^k \geq 0$

as a decision variable.

From the "Overall Concept S" representation we know that there are K such source regions.

Therefore, the algorithm generates x, as

$$x^T = (x^1, x^2, \ldots, x^K)$$

the vector of (crisp) decision variables.

Block 3
Using the conceptual constraints (described in Section 2 and represented in 1) and several auxiliary variables which are also defined as descriptors of slots of the concepts in the knowledge base, the algorithm related with 3 generates the core model using an analytic-type representation as follows:

$$y = Tx, \qquad \text{where } y^T = (y_1, y_2, \ldots, y_m) \tag{1}$$

$$T = (t_{j,k})_{m,K}$$

$x \geq 0$, non-negativity of decision variables

y grid-deposition-vector

$$\mathrm{Cost}(S) = \sum_{k=1}^{K} f_k(x^k) \qquad (2)$$

Block [7] Requirement Modelling

- Block [4]: The attributes (slot-names) which could be used to formulate user-defined constraints or goals could be selected as follows:

grid-deposition-of-grid-j:	y_j
level-of-emission-of-source-region-k:	x^k
cost-of-reducing-emission-to-level-x^k:	$f_k(x^k)$
cost-of-S:	Cost(S)

  All attributes listed above are assumed to be crisp.

- Block [5]: For selected subsets J of $\{1, ..., m\}$, K_1, K_2 of $\{1, ..., K\}$ the attributes related to y_j, x^k or $f_k(x^k)$ are chosen to be used as user-requirements-constraints.

- Block [5]: The cost-of-S attribute is selected to become a goal (objective function).

Block [7], [10], [8]:

In the MCDM-Model and Problem Knowledge Base there are predefined formulations for crisp constraints or crisp goals as well available to find the representation required in [8], e.g.:

If an attribute represented by a name z should be used to formulate an user-requirement-constraint, than there are the following alternatives predefined:

(α) $z_l \leq z$

(β) $z \leq z_u$

(γ) $z_l \leq z \leq z_u$

(δ) $z = \mathring{z}$

Of course, the alternatives (α) - (δ) require that values z_l, z_u, $\mathring{z}$ are already available or could be defined by the user.

Let us consider examples:

If the name-of-the-attribute equals to the level-of-emission-of-source-region, then one might use (γ) to establish the user requirement

$l_k \leq x^k \leq u_k$ for $k \in K_1$

and $x^k \geq 0$ otherwise.

One might easily remark, that if $z = y_j$ and

$y_j = \sum_k t_{kj} \bullet x^k$, ($\beta$) would be replaced using formula manipulation by

$t_{1j} x^1 + \ldots + t_{kj} x^k \leq (z_u =) b_j$ for any $j, j \in J$

The MCDM-model base also provides the user with possibilities to formulate goals (objectives).

If the attributes name chosen to act as a crisp goal is g, then there are the alternatives

(α')maximize g

(β')minimize g

(γ')define aspiration and/or reservation levels for g.

Finally, block [8] automatically generates an analytic representation of the so-called Requirement Model, e.g.:

One particular Requirement Model (crisp case)		
$l_k \leq x^k \leq u_k$	, for $k \in K_1$	(3)
$x^k \geq 0$,	, otherwise	(4)
$f_k(x^k) \leq c_k$	, for $k \in K_2$	(5)
$y = (Tx)_j \leq b_j$	, for $j \in J$	(6)
minimize Cost(S)		(7)

Figure 60: A special requirement model

The overall model is given by the composition of the core model with the analytic requirement model.

Now, one might want to assign different problems to this model, e.g.:

<u>Problem 1:</u> Find a feasible vector x (meaning that x fulfills (1), (3), (4), (5), (6)) such that Cost(S) becomes minimal, where Cost(S)

is defined by (2).

Problem 2: Find a feasible vector x ((1), (3), (4), (5), (6)) such that in addition an aspiration-reservation-level based constraint for Cost(S) is fulfilled instead of minimizing the Cost(S).

<u>Example 2</u> (Fuzzy Constraints)
For simplicity we assume that there is the same core-model as in Example 1 formulated by (1) and (2). We still have a crisp non-negative vector x of real-valued decision variables x^k. Also, the Requirement Model (block [7]) remains unchanged. However, in block [10] we use predefined formulations for fuzzy constraints, which means, we replace (α), (β), (γ) and (δ) by:

(α*) $z_l \lesssim z$

(β*) $z \lesssim z_u$

(γ*) $z_l \lesssim z \lesssim z_u$

(δ*) $z = \overset{o}{z}$, where $\overset{o}{z}$ is a fuzzy number.

(The relation $\lesssim$ we did introduce in Section 5.5.1)

As explained in Section 5.5.1, each of those constraints requires a predefined fuzzy set to model a smooth transition from full-feasibility to infeasibility. The respective fuzzy sets have to be associated with those slots of z, which have been selected (technically: fuzzy descriptors, fuzzy frames).

If we decide, for example, to use fuzzy constraints for all attributes we used in example 1, then we get:

(3') $l_k \lesssim x^k \lesssim u_k$ for $k \in K_1$

$x^k \geq 0$,

(4') $y_j \lesssim b_j$ for $j \in J$

(5') $f_k(x^k) \lesssim C_k$, for $k \in K_2$.

Because we have now fuzzy constraints (3'), (4'), (5'), we can not just require Cost(S) to be minimal as we did in Example 1. Following H.-J. Zimmermann's approach [53] we replace:

minimize-Cost(S)

by a fuzzy constraint

Cost(S) $\lesssim$ C-DES. (6')

C-DES (Desired Cost level) is a crisp bound which describes a cost level which is completely desirable. There are several approaches to determine C-DES and the related fuzzy set illustrated in Fig. 61.

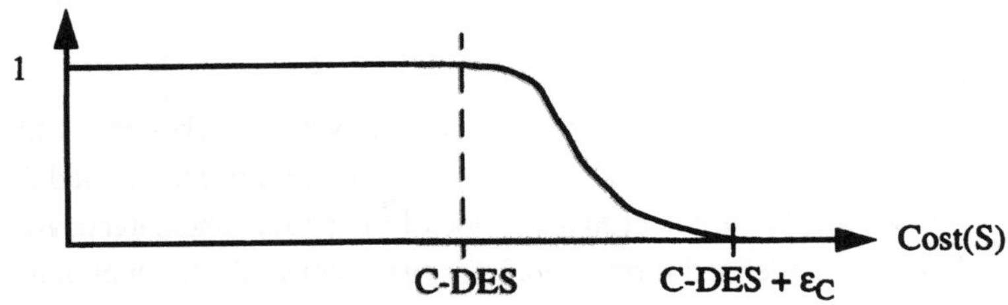

Figure 61: Fuzzy constraint "Desired Cost Level"

If the fuzzy set in Fig. 61 is determined, the question is how to formulate an optimization model and a related problem.

If we remember that each of the fuzzy constraints (3'), (4'), (5') and (6') is defined by a fuzzy set and that the constraints have to be fulfilled simultaneously, we can define a fuzzy set by applying the linguistic AND to all the fuzzy sets in (3') - (6'). More formally, if the fuzzy sets related to (3'), (4'), (5') and (6') are denoted by F_k, $\bar{F}_j$, F_k^* and $\tilde{F}_k$ respectively we get:

$$G_1 = \underset{k \in K_1}{\text{AND}}\, F_k, \quad G_2 = \underset{k \in K_2}{\text{AND}}\, F_k^*, \quad G_3 = \underset{j \in J}{\text{AND}}\, \bar{F}_j,$$

and $G = G_1$ AND G_2 AND G_3 AND $\tilde{F}$

All the "AND's" in the formulas above have to be interpreted by Fuzzy Operators as for example by L. Zadeh's Min-Operator or by several other operators including compensatory operators [55]. Then, an interesting modelling task is to choose the appropriate operator for each of the "AND's".

For example, let us consider the following aggregation tree in Fig. 62.

The AND-operators defining G_1, G_2 and G_3 should be modelled by the MIN-Operator, because "compensation" between countries related to reducing of emission or between grids related to deposition does not seem possible. However, an aggregation tree with γ-operators could be used to define the "AND" in the formula for G.

The aggregation of satisfaction with fulfilling the emission intervals on one hand and the satisfaction with fulfilling the cost-bounds on the other makes particular sense, because more realistically there is a fuzzy relation between x^k and $f_k(x^k)$ instead of a

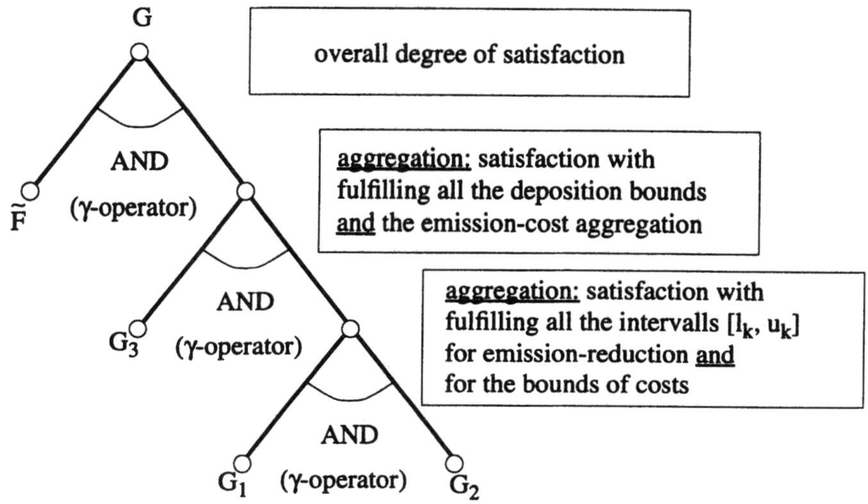

Figure 62: Aggregation tree

crisp function. Therefore, compensation also makes sense. Such a compensatory operator could be the γ-operator.

Finally, our problem is to maximize an "overall-degree of satisfaction". However, the solution we get is of course crisp not fuzzy. This means: We obtain a vector $\hat{x}$ which ensures a maximal degree of satisfaction of fulfilling all the constraints and goals finally modelled by the aggregation tree Fig. 62. For example, we can define $\hat{x}$ such that the membership function of the fuzzy set G will be maximized. If $\hat{x}$ is obtained, we can easily compute

$$\hat{y} = T\hat{x}$$

$$f_k(\hat{x}^k)$$

$$\mathrm{Cost}(\hat{S}) = \sum f_k(\hat{x}^k)$$

which are crisp values. Also, we can reconsider the fuzzy constraints to see where and how the solution $\hat{x}$ violates the constraints. In this sense, the solution is crisp rather than fuzzy, but it is elaborated by a fuzzy-based approach. There are several well known and efficient methods to solve such linear optimization problems.

Finally, we remark that the core model, in particular y = Tx could be replaced by fuzzy graphs to make the whole model closer to reality.

6 Summary and Conclusions

Chapter 3 is focused on configuration as a special class of engineering design problems.

Section 1 has an introductory character. We describe some general characteristics of Intelligent Decision Support Systems (IDSS). In addition, we introduce configuration as a special type of design tasks and we describe both routine and innovative design problems.

Furthermore in Section 1 we discuss the question "how to build an IDSS for Design and Configuration tasks" and we propose a general architecture for such a system. Finally, the role of Fuzzy Logic in IDSS for Design and Configuration tasks is discussed briefly within the introductory section.

Section 2 presents models and methodology to solve design and configuration tasks. We go much more deeply into the details than in Section 1, however we choose a modelling level which is still close to human understanding rather than understandable by computers.

In particular, we show, how Knowledge-based Systems might be used to model hierarchical knowledge, constraints and goals. We formulate our approach towards a Knowledge Model for configuration tasks, which has the particular advantage of dealing with imprecsion by using Fuzzy Logic. To illustrate the concepts we use in Section 2 very simple illustrative examples only, but later, in Sections 3 and 5 we apply our approaches to real world examples. In addition to the Knowledge Model we introduce a so - called Requirement Model which deals with the representation of user requirements. In particular, we focus to the Fuzzy Design Requirement Model(FDRM), which models imprecise user requirements using Fuzzy Logic. In particular, we present Fuzzy MADM, the Bellman-Zadeh, and some Defuzzification approaches in order to provide the reader with different alternatives to model Fuzzy Design Requirements. In Section 2.3 we present a model for optimization-based configuration using both crisp or fuzzy goals and crisp or fuzzy constraints as well. The optimization approach represented in this sub-section is very much influenced by techniques from the Artificial Intelligence area (tree search). In particular we develop a top-down approach with the aim to obtain the global optimum of a general configuration problem. The concepts of fuzziness are incorporated into these methods, which are completely new in the field of "Knowledge-based Configuration".

In subsection 2.4 a two-phase method is presented which deals with finding the global optimum of a special class of configuration problems.

The next three Sections 3,4, and 5 of this chapter are devoted to the application of the models and methods developed in Sections 1 and 2 and to some details of a computer-implementation.

In Section 3 we give a formulation of five completely different problems in order to show the application of our methods to Engineering Design problems, and also to configuration tasks from other fields. We restrict, within this Section, ourselves to brief formulations of the real world problems we want to deal with. We don't use any formalized representations of those problems, however the Knowledge Models of the applied problems will be introduced briefly. Three of our real world applications are Engineering Design or Configuration type problems:

- the passenger aircraft cabin layout application,
- the preliminary design and configuration of future Space Launch Systems problem, and
- the Personal Computer configuration problem.

The "Logistic Layout" application belongs more to the management science area and the example "Configuration of a Model in the Environmental Research Area" is an illustration of the use of configuration approaches to support the modelling process of an Optimization or Multi-Criteria Problem. The choosen application belongs to ecological modelling.

In Section 4 we introduce the KONWERK workbench, that means a kit of tools for configuration. KONWERK might be considered as an implementation of most of the concepts and methods developed within the German Research project PROKON and partly described in the Sections 1 and 2. KONWERK is implemented in LISP, running on Unix Workstations. It is public domain and can be used after contacting the authors. Of course, Section 4 can not provide the reader with all details of the KONWERK syntax. However, persons who are experienced with FRAME-like languages should have no problem interpreting the pieces of syntax we are using in this Section.

One particular focus of Section 4 is to introduce the concepts of Fuzziness which are currently available in KONWERK. Similarily to Section 2 we use simple illustrative examples and we use the real syntax, which should be understandable because we added all the necessary explanations.

KONWERK (which is still a prototype) might be considered to be the first available workbench for knowledge-based configuration of complex systems which has features of "optimization" and "dealing with fuzziness".

In Section 5 we reconsider the applications introduced in Section 3 to show how to solve them using either KONWERK or some other methods or tools. The degree of detailed consideration is different within this Section. We hope the reader understands that it is impossible to show the complete solutions of the five applications (the whole software systems, etc). Therefore, we picked for each of the examples one or two particular points of interest from the complete solution process to give a feeling about the problem solving process and the solutions as well. For the applications in Subsections 5.2 and 5.4 the complete solution is available (in German language). For the other examples the solution (software) is partly documented. For all five examples see the respective references.

In Section 5.1. we illustrate - for the aircraft cabin layout problem the two-phase approach introduced in Section 2.4. A small set of "good" solutions (minimization of costs) is constructed by a Mixed Integer Problem solver. In the second phase a Fuzzy -MADM is used to rank the good layouts using four fuzzy goals.

In Section 5.2 we illustrate two aspects of the Space Launch System example. We illustrate parts of the domain knowledge base, which is completely implemented in KONWERK, and we show the goal hierarchy of the AHP and examples of pairwise comparisons. Also the Logistic Layout problem is completely implemented using the KONWERK workbench with a nice interface to visualize the used geographic data. The example "PC-configuration" using evolutionary algorithms was an experiment to approach the global optimum of a multi-criterial configuration problem with imprecise and crisp goals as well.

Meanwhile, our implementation shows interesting results. Therefore, we don't focus on the genetic operators but on the multi-criteria-evalution method of populations of PC's which allows a hierarchy of crisp and fuzzy goals as well. For the application described in Section 5.3 there is no implementation until now.

Summarizing, this chapter presents three topics: modelling, software tools and real world applications devoted to configuration problems. It combines intelligent techniques as for example Knowledge-based techniques, Fuzzy Logic, Multi-criteria Decision Making, Evolutionary Algorithms and others in order to solve real world problems. The fuzziness in our configuration model is mostly restricted to user formulated requirements, where the user also might be a designer. However, the design artifacts are crisp objects. Therefore, a very interesting generalization will be to merge the "method of imprecision" presented in Chapter 1, with the methods described here.

References

[1] Agogino, A. M.; Bradley, S. R.; Cagan, J.; Jain, P.; Michelena, N., 1989: AI/OR Computational Model for Integrating Qualitative and Quantitative Design Methods; Proceedings of NSF Engineering Design Research Conference, Amhurst, June 11-14, pp. 97 - 102

[2] Alcamo, J.; Hordijk, L.; Shaw, R.: The RAINS model of acidification: Science and Strategies in EUROPE, IIASA/Kluwer Academic Publisher, Dordrecht, 1990

[3] Altman, A.; Amann, M.; Klaasen, G.; Ruszczynski, A.; Schöpp, W.: Cost-Effective Sulphur Emission Reduction Under Uncertainty, IIASA Working Paper, WP-94-119, Nov. 1994

[4] Appelrath, H.-J., Sebastian, H.-J., 1992: Planen und Konfigurieren: eine integrierte Modellbildung; PROKON-Memo 2

[5] Chen, S.-J.; Hwang, C.-L.: Fuzzy Multiple Attribute Decision Making; Springer, Berlin, 1992

[6] Cunis, R.; Günter, A.; Strecker, H., 1991: Das PLAKON-Buch; Informatik Fachberichte 266; Springer, Heidelberg

[7] Fischer, G. W.; Antoine, J.: Agro-Ecological Land Resources Assessment for Agricultural Development Planning, A Case Study of Kenya, World Soil Resources Reports 71/9, IIASA, FAO, Rome 1994

[8] Funke, B.: Das Optimierungs-Grundmodul - Beschreibung und Einbindung in KONWERK; PROKON-Memo Nr. 53, Juni 1994, RWTH Aachen

[9] Funke, B.: The Optimization Modul of KONWERK - Using Methods of Operations Research Within Knowledge Based Configuration of Systems; 2.'nd EUFIT, Aachen, Sept. 20 - 23, 1994

[10] Funke, B.; Müller, K.; Sebastian, H.-J.; Thärigen, M., 1994: Innovative Design by Integration of Expert System Technology with Multi-Criteria Decision Making and Fuzzy Systems; Second World Congress on Expert Systems, Lisbon, Proceedings

[11] Geoffrion, A. M.: An Introduction to Structured Modelling, Management Science, Vol. 33, No. 5, 1987

[12] Gero, J. S., Maher, M. L., Zhao, F., 1988: A Model for Knowledge-Based Creative Design; Proceedings of AAAI-88 Workshop on AI in Design, St. Paul

[13] Goudarzi-Pour, T.: Wissensrepräsentation in Wissensbasierten Systemen und Anwendung im Bereich Konfigurierung; Diplomarbeit, RWTH Aachen, 1995

[14] Günter, A., Dörner, H., Gläser, H., Neumann, B., Posthoff, C., Sebastian, H.-J., 1991: Das Projekt PROKON: Problemspezifische Werkzeuge für die wissensbasierte Konfigurierung; PROKON-Bericht 1

[15] Günter, A.: KONWERK - ein modulares Konfigurierungswerkzeug, in F. Maurer and M. M. Richter (Hrsg.), Expertensysteme 95, Kaiserslautern (1995).

[16] Günter, A.: Modelle beim Konfigurieren, in O. Herzog, T. Christaller and D.\x11Schütt (Hrsg.), 17. Fachtagung für Künstliche Intelligenz Grundlagen und Anwendungen der Künstlichen Intelligenz, Springer Verlag, Berlin, 1993, pp 169-176.

[17] Günter, A.; Sebastian, H.-J.: The KONWERK-Workbench: A System of Software Tools to Approach Intelligent Design Support, The Second EUFIT '94, Proceedings, Aachen, 1994

[18] Günter, A. (ed.): Wissensbasiertes Konfigurieren: Ergebnisse aus dem Projekt PROKON; INFIX, Sankt Augustin, 1995

[19] Harmon, P.: The Market for Expert Systems-Buildung Tools, in Intelligent Software Strategies, 9 (2), pp 3-9., 1993

[20] Järvinen, S.: Optimierung bei der Konfigurierung, PROKON-Memo Nr. 36, Aachen, 1993

[21] Järvinen, S.: Optimierungsbasierte Konfigurierung in intelligenten Decision Support Systemen, Lizentiatarbeit, Aachen und Helsinki, 1994

[22] Kopisch, M. and Günter, A.: Configuration of a passenger aircraft cabin based on conceptual hierarchy, constraints and flexible control, in F. Belli and F. J. Radermacher, (ed.), 5th International Conference Industrial and Engineering Applications of Artificial Intelligence and Expert Systems, IEA/AIE-92 (Springer, Paderborn, 1992), pp 421-430.

[23] Günter, A.; Kopisch, M. and Sebastian, H.-J.: Integration of Knowledge-based Configuration with Fuzzy Logic and Optimization, in: Hirota, K. (ed.): Advances of Fuzzy Logic, Vol. 2: Industrial Applications of Fuzzy Logic in the World, World Scientific Publishers, Singapore, 1995

[24] Kopisch, M.: Spatial relations for configuration tasks in general and for the cabin layout of the AIRBUS A340 in particular, in P. W. H. Chung, G. Lovegrove and M. Ali, (ed.), 6th International Conference Industrial and Engineering Applications of Artificial Intelligence and Expert Systems, IEA/ AIE-931, Gordon and Breach Science Publishers, Edinburgh, 1993, pp 450-457.

[25] Kriese, T.: PC-Konfigurierung mit evolutionären Algorithmen, Diplomarbeit, RWTH-Aachen, 1996

[26] Laporte, G.; Osman, I. H.: Routing Problems: A Bibliography, CRT-94-52, Montreal, 1994

[27] Laporte, G.: The Vehicle Routing Problem: An overview of exact and approximate algorithms, European Journal of Operational Research 59, 1992

[28] Lötzerich, K. and Goudarzi-Pour, T.: Knowledge-Based Configuration of future Space Launch Systems, in EUFIT 94, Proceedings, 1994.

[29] Lötzerich, K., Sebastian, H.-J. and Goudarzi-Pour, T.: Multi-Criteria Decision Making (MCDM) for rational evaluation and selection of future space launch Systems, research report, 1994

[30] Makowski, M.: Methodology and a Modular Tool for Muliple-Criteria Analysis of LP-Models, IIASA Working-Paper WP-94-102, Dec. 1994

[31] Mittal, S. and Frayman, F.: Towards a generic model of configuration tasks, in Proceedings IJCAI-89, 1989, pp 1395 - 1401.

[32] Müller, K. and Thärigen, M.: Requirement modelling imprecise knowledge and Multiple Criteria Decision Making in KONWERK - a way towards an intelligent Decision Support System, in 2. European Congress on Fuzzy and Intelligent Technologies, Aachen, Sept. 20 - 23, 1994

[33] Müller, K.: Fuzziness in KONWERK, EUFIT 95, Proceedings, 1995

[34] Navinchandra, D.: Exploration and Innovation in Design, Springer Verlag, 1991

[35] Otto, K. N.: A formal representational theory for engineering design, Ph. D. Thesis, California Institute of Technology, Pasadena, 1992

[36] Otto, K. and Antonsson, V. E.: Modelling imprecision in product design, in Proceeding of the 3'd Fuzzy-IEEE, Orlando, 1994

[37] Rao, S. S. and Dhingra, A. K.: Applications of fuzzy theories to multi-objective system optimization, NASA Contractor Report 177573, NASA Ames Research Center, Moffett Field, CA 94035-1000, University Consortium Interchange Number CA2-223, 1991

[38] Rao, S. S. et al., Multiobjective fuzzy optimization techniques for engineering design, in Computers and Structures, 42 (1), pp 37 - 44, 1992

[39] Sebastian, H.-J. and Dilger, W.: Knowledge-based configuration of technical systems including optimization-based reasoning, in Proceedings of the 1'st World Congress on Expert Systems, Pergamon Press 3, pp 1967 - 1974, 1991

[40] Sebastian, H.-J.: Integration of Knowledge-based Methods, Fuzzy Logic and Multi-Criteria Decision Making to Support Modelling, IIASA Working Paper, Laxemburg, 1995

[41] Sebastian, H.-J.: Intelligente Entscheidungsunterstützung bei der Systemkonfigurierung durch Integration von Expertensystemtechnologie mit Modellen und Methoden des Operations Research, PROKON-Memo Nr. 47, Aachen, 1994

[42] Sebastian, H.-J.: Intelligente Entscheidungsunterstützung bei der Systemkonfigurierung, in: Werners, B.; Gabriel, R. (eds.): Operations Research. Festschrift zum 60. Geburtstag von H.-J. Zimmermann, pp. 275 ff., Springer Verlag, 1994

[43] Sebastian, H.-J.: Modellierungsunterstützung - Ein Ansatz zur interaktiven Bestimmung eines problembezogenen Konfigurierungsmodells, PROKON-Memo Nr. 31, Aachen, 1993

[44] Sebastian, H.-J.: Optimierung und Unschärfe bei Konfigurierungsproblemen, PROKON-Memo Nr. 12, Aachen (1992).

[45] Sebastian, H.-J.: Optimierungsbasierte Kontrollstrategien beim Konfigurieren, PROKON-Memo Nr. 49, Aachen (1993).

[46] Sebastian, H.-J.: Optimization-based control-strategies in expert systems for design and configuration, in Proceedings of IFIP 7. Working Conference on Optimization-based Computer-aided Modeling and Design, Prague (Lansa-BV, 1994).

[47] Thärigen, M.: Realisierung der Anforderungsmodellierung in KONWERK, PROKON-Memo Nr. 50, Aachen (1994).

[48] Thärigen, M.: Modellierung von Konfigurierungsaufgaben, PROKON-Memo Nr. 25, Aachen (1993).

[49] Tong, C. and Sriram, D. (ed.), Design representation and models of routine design, (Academic Press, 1992)

[50] Zadeh, L. A.: Fuzzy Sets, in: Information and Control 8 (1965), pp. 338-353

[51] Zimmermann, H.-J. and Sebastian, H.-J.: Fuzzy design - integration of fuzzy theory with knowledge-based system design, in Proceeding of the World Congress on Computational Intelligence, invited paper Fuzzy-IEEE 94, Proceedings, Orlando, Florida (1994).

[52] Zimmermann, H.-J., and Gutsche, 1991: Multi-Criteria Decision Making; Springer, Heidelberg

[53] Zimmermann, H.-J.: Fuzzy Programming and Linear Programming with Several Objective Functions, Fuzzy Sets and Systems 1, pp 45 - 55, 1978

[54] Zimmermann, H.-J.: Fuzzy Set Theory - and Its Applications, Kluwer, Nijhoff Pub. Boston, Dordrecht, Lancaster, 1984

[55] Zimmermann, H.-J.: Fuzzy Sets, decision Making and Expert Systems; Kluwer Acad. Pub. Boston, Dordrecht, Lancaster, 1989

[56] Zimmermann, H.-J.; Sebastian, H.-J.: Intelligent System Design Support by Fuzzy Multi-Criteria Decision Making and Evolutionary Algorithms, FUZZ-IEEE 95, Proceedings, Yokohama, March 1995

[57] Zimmermann, H.-J.; Sebastian, H.-J.: Optimization and Fuzziness in Problems of Design and Configuration, Proceeding of the IEEE-Conference on Fuzzy-Systems, San Francisco, 1993

4

MANAGEMENT OF UNCERTAIN INFORMATION IN SIMULTANEOUS ENGINEERING

Professor Dr.-Ing. Dr. h.c. Walter Eversheim*

Professor Dr. Dr. h.c. Hans-Jürgen Zimmermann**

Dipl.-Ing. Dipl.-Wirt. Ing. Axel Roggatz*

Dipl.-Ing. Dipl.-Wirt. Ing. Thomas Derichs**

** Laboratory for Machine Tools and Production Engineering,*
*** Institute of Operations Research,*
RWTH Aachen, University of Technology, 52056 Aachen, Germany

ABSTRACT

The basic idea of Simultaneous Engineering is to parallelize activities in order to reduce the development lead time and thus the time to market. The processing of incomplete and uncertain information can achieve an enormous advantage in comparison to parallelization of activities based on exact information alone. Thus many activities can be started even if its predecessor is not completed. This article focuses mainly on two concepts to provide adequate information management for Simultaneous Engineering. The first concept bases on a systematic integration of uncertainty in the information flow. The second concept allows the break-up of the sequential and document-oriented information flow by creating small information units as needed. The application of both concepts supports an early information flow. Thus it becomes possible to coordinate activities more efficiently because critical topics can be recognized and balanced in advance, shifting the communication requirement to the early stages of the development process. Hence, expensive iterations, especially in the late stages of the product development process, can be avoided and the length of the iteration loop can be reduced.

1 INTRODUCTION

Product life cycles and pay-off periods have been steadily reduced in the last years. In combination with a growing product complexity it becomes harder and harder for a conventional product development process to fulfill the rapidly changing requirements. Nowadays price and quality are no longer the only relevant criteria which guarantee the economic success of a product. The right timing of a product introduction gains in importance and this usually means that the introduction should take place as early as possible [6; 10; 11].

A major reason why enterprises react slowly and inflexibly to new demands is the sequential process of product development [5]. Usually departments act more or less independently, not knowing the demands and the capabilities of each other. In general, commercial information and communication systems support only a sequential work flow but no parallelism. They are designed to handle documents and exact information. Neither partial information (information units), nor the transmission of uncertain, fuzzy and incomplete information are supported by these systems, even if this type of data contains important and valuable information for succeeding activities.

Particularly in the early stage, uncertain and incomplete information is typical for the product and process development. This information contains a lot of constraints and determines the main part of the development process.

The goal of this article is to present a methodology which enables the employees of an enterprise to provide and use partial and uncertain information in a very early stage of the product and process development. The benefit of this additional information lies in its capability to decrease the number of iterations to reduce the time between creating and using information.

The different aspects which must be considered when uncertain information is used will be presented in section 2. Special requirements concerning the contents and the context in which information has to be regarded will be presented in section 3 and 4. Section 5 deals with the modeling and information architecture of context dependent uncertainties. The following section shows a way to structure and subdivide classical documents by using an object-oriented schema into information units to achieve an acceleration in the information flow. The last section gives an example of how the concepts "integration of uncertain information" and "usage of small information units" fit together and how they can solve real-world problems.

2 ASPECTS OF MODELING UNCERTAIN INFORMATION

The essence of any information is generally determined by two components. Both of them can contain uncertainties. The first component describes the information-content itself. The second component contains additional information about the reliability of the contents and describes the context in which the information has to be viewed. For modeling uncertain information it is necessary to handle both sources of uncertainty.

Typical reasons for uncertainties concerning the contents of information are imprecision and incompleteness or vagueness. Linguistic variables are an appropriate tool for their modeling. This concept was developed in Fuzzy Set Theory [12, 13]. They assign a semantic meaning to verbal expressions. The values of linguistic variables are not numbers but words or sentences which are called terms. Linguistic variables are a powerful tool to reduce complexity because they aggregate the subjective estimation of the described facts.

The most significant reason for misinterpretation of information is the loss of context-information. The recipient of information uses his knowledge about the context to interpret the statement. If he misses the real context he has no possibility to judge the reliability. Hence, he has to guess and is faced with an uncertainty of the whole statement and not of single values. Both components, the contents and the context, affect the essence of information which can be extracted by the recipient. Hence, for both components a concept to model uncertain information is required (figure 1).

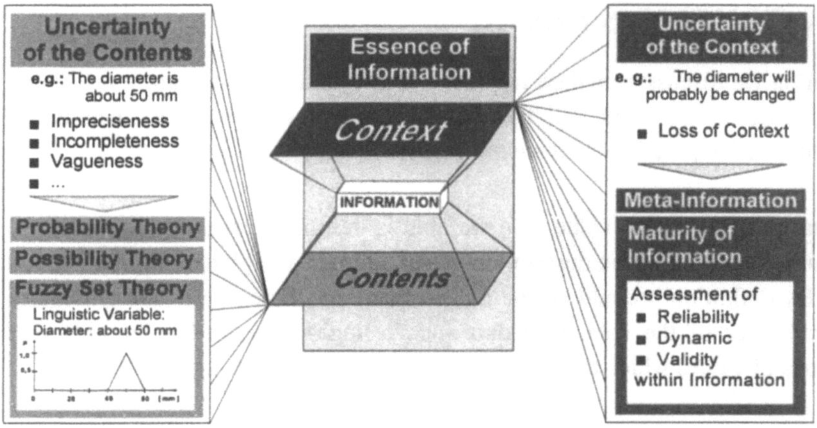

Figure 1 Aspects for Modeling Uncertain Information

- The content of information is related to the data level. It contains uncertainties, which refer to data and their values. The data at this level are entirely defined but their values contain an uncertainty. In this case the possibility to present vague data is required. An example of a defined but vague data item is within the statement: *'The diameter of the shaft X is about 50 mm'.*

- The context of information concerns the conditions, the immediate context, of data. The uncertainties at this level refer to the subjectivity and reliability of given values at the data level. Hence, it is necessary to develop a method which supports the interpretation of data. Through this, the data can be handled as facts. Concepts are needed to allow the evaluation of the reliability of information and to take the subjectivity into consideration. To interpret the statement *"Presumably the diameter of the cylinder is 30 mm"*, knowledge about the context is needed. In this example, the context contains the knowledge of the person who is the source of the information. The context is needed to assess information.

To develop a methodology to present and to provide uncertainties, it is necessary to find solutions for both aspects of uncertain information. The integration of these aspects allows us to define uncertain information and creates the basis to parallelize the information flow within the product and process development.

3 MODEL FOR A DESCRIPTION OF UNCERTAINTIES ON A CONTENTS LEVEL

In this section we discuss a method to present uncertainties on the data level. Thus, uncertainties of the contents will be examined, but not uncertainties within the context which are required to evaluate information.

At the data level the actual value of a variable is described primarily by numbers. The semantic definition of the description is well defined and given. A value is crisp or vague. By definition it could not be multi-valued. Multiple values and ambiguity are characteristics of the data. If a data item is multiple valued it has different instantiations of values which themselves could be crisp or vague again.

Fuzzy set theory offers the possibility to describe vagueness. This theory is a generalization from classical set theory. It is developed for solving problems in which descriptions of activities and observations are imprecise, vague, and uncertain. A fuzzy set is a class of objects with a continuum of membership grades. A membership function, which assigns to each object a grade of membership, is associated with each fuzzy set. Usually, the membership grades are in [0,1]. When the grade of membership for an object in a set is one, this object is absolutely in that set; when the grade of membership is zero, the object is absolutely not in that set. Borderline cases are assigned numbers between zero and one [2].

To represent the data model an object-oriented method has been chosen. This method allows the integration of a process and a data view [7]. It allows also the connection to the international standard STEP (Standard for the Exchange of Product Model Data). This standard uses EXPRESS, which is an object-oriented modeling language [9]. In addition, STEP contains different models that describe the main areas of the product and process development.

The integration of linguistic variables in an object-oriented model for the representation of vagueness is an effective combination to:

- handle complex data structures,
- further expansions of the modeling and
- include uncertainties in modeling.

The model has to contain certain and uncertain data in a heterogeneous description. In particular the following values need to be represented:

- linguistic values (large, middle, small,...),
- crisp numeric values (7. 5, 19, . . .),
- uncertain numeric values (around 75, about 17 to 19) and
- alternative values (either 4 or 5).

Crisp values are represented in all existing data models. Also, alternate values can be described in object-oriented models by using different instances of the same attribute. To manage these different value types, special methods within the objects have to be implemented. To do this, a model to describe linguistic and uncertain numeric values is needed. To support the model, a new data type 'FuzzyObject' is developed. This data type is an extension of the elementary data types represented in EXPRESS.

The object-oriented paradigm allows the addition of data types in existing models. The data types are available for all objects when they are implemented. In this manner, it is possible to expand existing models without changing their internal structure. The basic structure of the data type 'FuzzyObject' is shown in figure 2. It allows the presentation of fuzzy sets and the use of linguistic variables. These linguistic variables can be described also by fuzzy or crisp values.

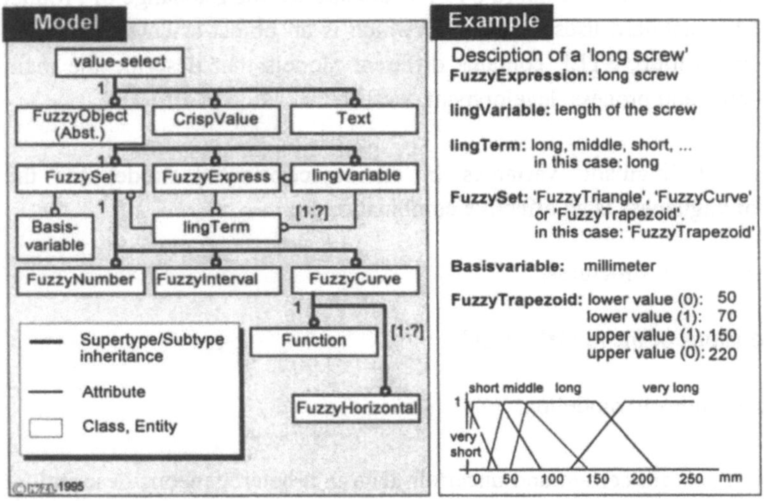

Figure 2 Presentation of uncertainties on a contents level described in EXPRESS-G

4 METHODOLOGY FOR A CONTEXT DEPENDENT DESCRIPTION OF UNCERTAINTIES

The interpretation of information contents is usually done by considering as far as possible all available context information. The most important reason for mis-interpretations is the loss of context-information. Usually documents have no place to specify such information. Because it is hardly possible to take into account all context information, it is necessary to focus the context information on the most expressive information. It is the objective of this work to provide meta-information about the certainty, reliability and its dynamic development accessible to each who receives uncertain information. The relevant information about the context already exists in the mind of the information provider. In a verbal communication a person has the possibility to absorb most of this meta-information. When information is stored in databases or aggregated in documents this information gets lost. To provide this additional information, a concept was developed which is called "information-maturity".

In analogy to nature one can observe a "process of maturing" for information in product development. From the first product idea to a saleable product, all information becomes more precise, reliable and certain. This can be summarized as an improvement in maturity. This basic idea was used to describe the status of a piece of information. Information-maturity facilitates an adequate interpretation of the reliability, variability and validity of an information contents. The information-maturity reflects the judgment of the information provider enabling a recipient to assess this information concerning its usability for his current task. In this manner, an additional potential for a parallelism in product and process design can be achieved.

4.1 CRITERIA for Determining "information-maturity"

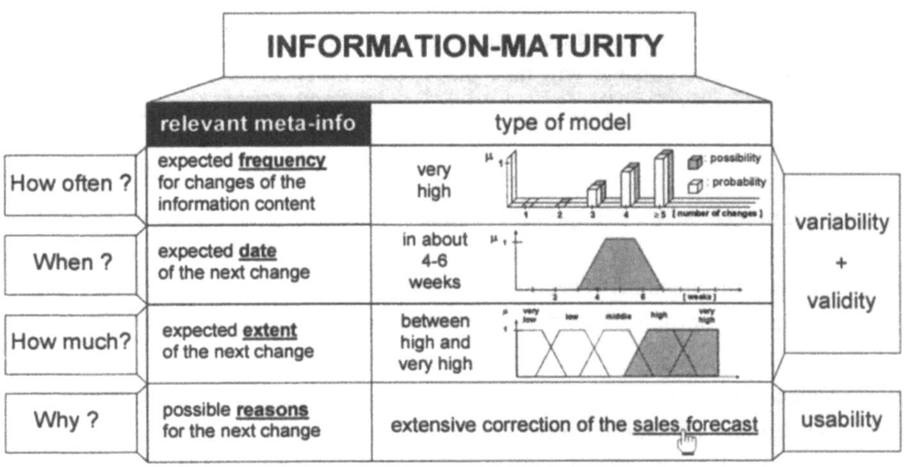

Figure 3 Determining the information-maturity

To determine the maturity of information it is necessary to select a level of aggregation that provides a balanced compromise between the expense of acquiring additional relevant information and its benefits due to unequivocal interpretation. The use of information-maturity is necessary where the provider of information judges it as relevant for parallelism. In the context of maturity, uncertainty indicates that the contents of the information will change in the future at least one time. Relevant for parallelism means that costly iterations can be avoided by making information early available.

Seeking an adequate method to model information-maturity, it seems reasonable to use the notion of a degree of maturity, similar to certainty factors in artificial intelligence. A close examination of this shows, however, that this is not sufficient. The prime reason is the high level of aggregation. On the one hand, the assignment of degrees for the specification of uncertainty is not unequivocal and depends on a subjective judgment of the situation; on the other hand, its interpretation is ambiguous. Moreover, it is likely that acceptance problems will occur, especially with small degrees of maturity.

To avoid these problems an approach has been developed which specifies four attributes to determine the information-maturity (figure 3).

The dynamic aspects and the variability of an unit of information can be determined by asking the information provider how often the contents of the information will change in the future. The next three questions relate only to the next change. They express when the next change will take place, how much the contents will change from the present to the next stage, and what are the most important reasons for the change.

The specification of the most important reasons allows the quick identification of the responsible person, to whom one has to turn to get more detailed information. Through these four questions it becomes possible for the information recipient to get a better impression of the validity and the usability of the uncertain information.

The representation of the first three criteria is based on fuzzy set theory. It allows a flexible description of data and gives the possibility to define subtle nuances. The fourth criterion will be expressed linguistically.

Frequency of changes

For the description of the frequency of change a possibility distribution and probability distribution have been selected. The possibility is an upper bound for the probability [3]. This theory supports the user by specifying his estimate concerning the frequency of change. First, the user has to select one out of five linguistic terms (very low, low middle, high, very high), which describe the possibility of changes (table 1). In the next step the user has to estimate subjective probabilities for each possible event. This can be facilitated, for example by using graphical scales. The values of this indirect measurement will be transformed into subjective probabilities.

Table 1 Linguistic terms for a description of the changing frequency

	Number of Changes				
	1	2	3	4	≥ 5
Linguistic Term	Degree of Possibility				
very high	0,00	0,00	0,33	0,66	1,00
quite high	0,00	0,33	0,66	1,00	0,66
middle	0,33	0,66	1,00	0,66	0,33
quite low	0,66	1,00	0,66	0,33	0,00
very low	1,00	0,66	0,33	0,00	0,00

Date of change

Specifying the exact date for a change is usually quite difficult. Hence it seems more appropriate to express this statement linguistically, e.g. "the change will take place in approximately 3 weeks". Conventional databases have no possibility to model such an information. With the availability of linguistic variables it becomes possible to model and to handle even uncertain temporal data. Very often piecewise linear functions are chosen to describe the term semantics. Triangular or trapezoidal membership functions can easily be specified by the user and are computationally efficient [12]. A user has to answer two questions, if he wants to find the membership function for the term "expected date of change in about 3 - 4 weeks":

1) Which interval belongs certainly to "about 3 - 4 weeks"?
The interval [3, 4] could be an answer. It has a membership degree of $\mu = 1$.

2) Outside of which interval is "about 3 - 4 weeks" certainly wrong?
The interval [2, 5] could be an answer. It has a membership degree of $\mu = 0$.

With linguistic variables the information provider has the possibility to assign a fuzzy interval to a date of change. Terms like 'approximately' or 'about' can be equipped with user-specific semantics.

Extent of change

Typical instruments to extract information like "the extent of change" are scores. The use of linguistic variables offers, in comparison to this instrument, a still higher flexibility. It allows the user to differentiate by fuzzy intervals between several items.

For "the extent of change" the extrema are 0 and 1. They represent the opposite judgments of a situation, like 'very low' and 'very high'. The linguistic variable "extent of change" has five terms (very low, low, middle, high, very high). By the number of terms and the shape of the membership functions the semantic can be defined.

Reasons for a change

The first three criteria allow a fast estimation of the information concerning its variability and validity. For a profound judgment of the situation it is often necessary to know further details. Via these details the recipient gets additional information. The application of hyperlinks and hypertext ensures a fast navigation even through distributed data base structures. Thus it is easy to find further relevant information units in documents like drawings, work plans, etc. Another important fact is the direct identification of the employee who is responsible for the change. This early contact can avoid iterations because demands and requirements are much earlier identified.

5 INFORMATION MODEL TO DESCRIBE UNCERTAINTIES WITHIN THE CONTEXT

In this chapter the presentation of uncertainties at the context level will be shown. In contrast to the developed data type 'FuzzyObject' which is discussed in section 3, the uncertainty at the context level is characterized by meta-information. The meta-information contains the information 'frequency', 'date', 'extent' and 'reason' which are introduced above. It is not possible to foresee for which objects meta-information is required. Thus, a generic model has to be developed which describes the meta-information. On the other hand it has to support an a posteriori integration in existing models. Moreover the extension of each object in these models must be possible. In such a case it makes no sense to develop a model which is treated as an attribute to the existing objects. In that case, all objects would have to be expanded, entailing an enormous effort.

Hence, an alternative was found by using the inheritance mechanism within the object-oriented paradigm. The inheritance mechanism allows the transfer of characteristics of one object to another using an inheritance hierarchy. This allows us to build up a model containing the description of meta-information and propagate this structure to the whole existing data model.

The developed model describes the four meta-information attributes discussed in section 4. In addition, a connection to the data type 'FuzzyObject' (figure 2) is necessary because the values of the context information are probably uncertain and vague. The core of the context model is the object "MetaInformation". This object has attributes which present the different descriptions of the context information as described above. By inheriting the object 'MetaInformation' to all other objects of the model, the defined attributes are available to them. To present the context information for an instantiation, the inherited attributes characterize their description. The extension of existing models by one object, whose behavior and attributes are inherited by the other objects, allows an a posteriori integration.

With the help of the mentioned models, uncertainties at the contents level (section 3), as well as uncertainties at the context level can be described and provided. To support a specific and demand-oriented pass on of information, further mechanisms are necessary. One of them is the flexible definition of small information units. Only if the uncertain information can pass on very early can the necessity of discussion and decision be recognized and the possibility of a parallization be exploited.

6 EARLY PASS ON OF INFORMATION BY STRUCTURING INFORMATION

In today's processes, classical document structure is a restriction for the parallization of activities. Generally most of the information is collected in documents and passes on before the document is completed. Normally, the recipients of a document require a fraction of the information collected in a document. But there is no attention paid to the need of small pieces of information in following activities. By using small and flexible information units, information can pass on and be used very early. Hence, a feedback to assess results and to actualize knowledge bases can be advanced. An information unit is a minimum logical unit of data, with which to present information. That means it is interpretable by itself. There are different possibilities to structure information in this context.

- Information units assigned to functions
  Combining input information of business functions into information units allows the assessment of the input depending on the actual situation and thereby information and the specific decision to start a function. In addition, it is possible to search for missing information.

- Information groups to conclude complex interconnection of information
  The conclusion of complex interconnection within the information allows a structure by topics. This structure also contains classical documents. Information groups model views of the information and facilitate the search for special contents.

- Functional relations between information
  Functional relations between information provide the recognition of logical and functional connections between information. With this, it is possible to estimate the effect of decisions because they allow us to identify topics related to the decisions.

The previously mentioned possibilities to structure information support different characteristics to accelerate the passage of information. These are a process-oriented, a contents-related and a decision-oriented structure. In application, these possibilities have to be combined not only with each other but also with the information model. In figure 4 an object-oriented schema modeled in EXPRESS-G is shown.

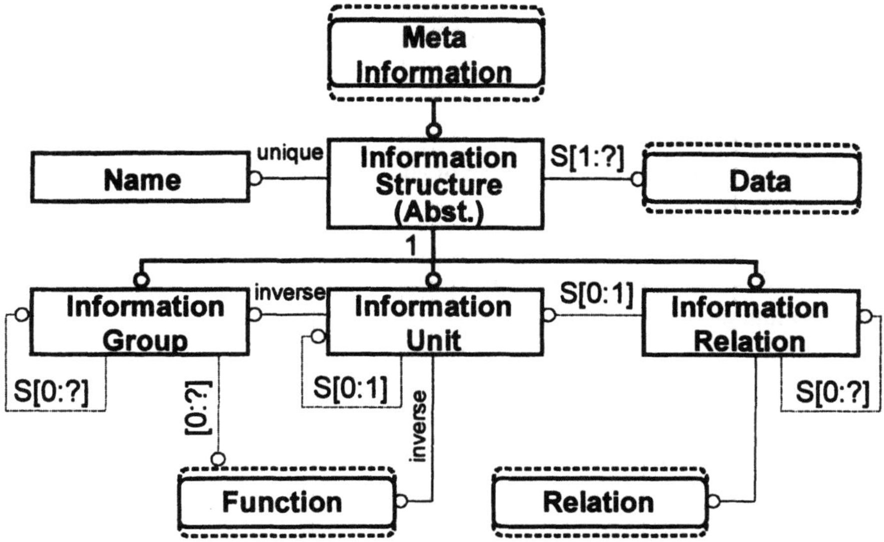

Figure 4 EXPRESS-G Model for structuring information using an object-oriented schema

The division of documents into small information units can be demonstrated through the example of a drawing. Normally this drawing has to become standardized and reviewed before it is given to other departments. But often later activities (e. g. the selection of a machine tool) just require a little approximate information. In such a case, for example, it can be sufficient to know just the rough dimension of the part from this document.

On first view it seems that using uncertain information and small information units will increase the complexity and the volume of communication. By developing an application to handle this kind of information, it can be shown that the effort of the user can be decreased and the complexity of the information can often be reduced.

7 APPLICATION OF THE METHODOLOGY

The use of the developed methodology requires EDP-support. It must be guaranteed that the four criteria of information-maturity can be fed into the system in a user-friendly way. Also it has to be ensured that relevant uncertain information can be found promptly and easily.

The query process is divided into three steps:

1st step: Selection of a search area to find uncertain information.

2nd step: Formulation of a fuzzy query.

3rd step: User-specific ordered output of all uncertain information from the search area.

In the first step, the user has the possibility to combine four different search strategies to find all relevant uncertain information which he needs to fulfill a certain task. Each information unit, which is created or modified, contains information concerning:

- the **activity** to which it refers,
- the **person** who created or modified the information unit and his position in the company organization,
- the **product** it is associated with, and,
- the **document** to which an information unit belongs.

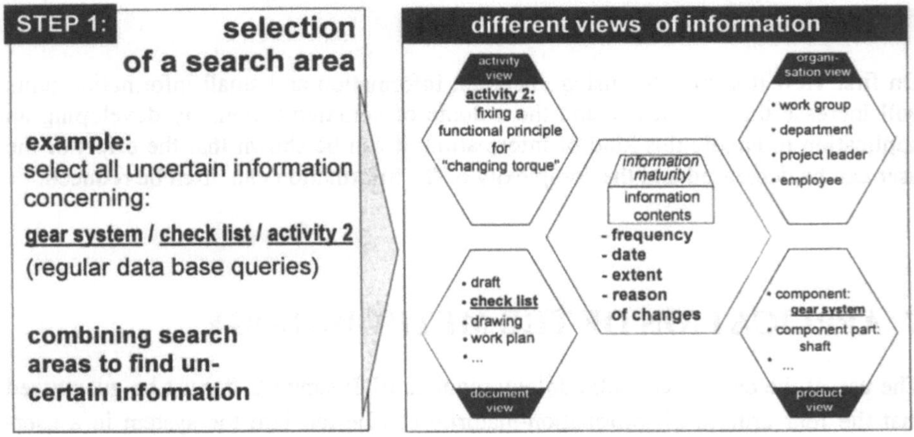

Figure 5 First step for the use of information-maturity in Simultaneous Engineering

The system allows us to focus on specific different activities in the product and process design. It checks whether the activities are linked to uncertain information or not. By means of this **activity view** all uncertain information, which are attached to an activity, can be filtered out. Using the **organization view**, the search procedure allows us, for example, to find all uncertain information which was provided by a certain work group or in a certain department. The **product view** enables the user to identify uncertain information which belongs to products, components or component parts. The purpose of the **document view** is to detect uncertain information units in documents, like check lists, drawings or work plans. Combining these four views of information, the search area can be effectively limited to find uncertain information. Figure 5 shows the first step that combines the four views.

The first step of this procedure can be realized by conventional data base operation, whereas the second and the third steps require a more enhanced technology to enable an adequate exploitation of the meta-information. The system now provides the means to specify linguistic queries. This enables the user the possibility, for example, to ask for "all information which will change in approximately six months or later". Similarly it is possible to ask for "all critical information". This linguistic query requires that the user-specific and personal estimation of "critical" has to be transformed into the first three meta-information (frequency, date and extent of changes). One "translation" of "critical" could be a very high frequency of changes in combination with a late and at least a high extent of the expected change.

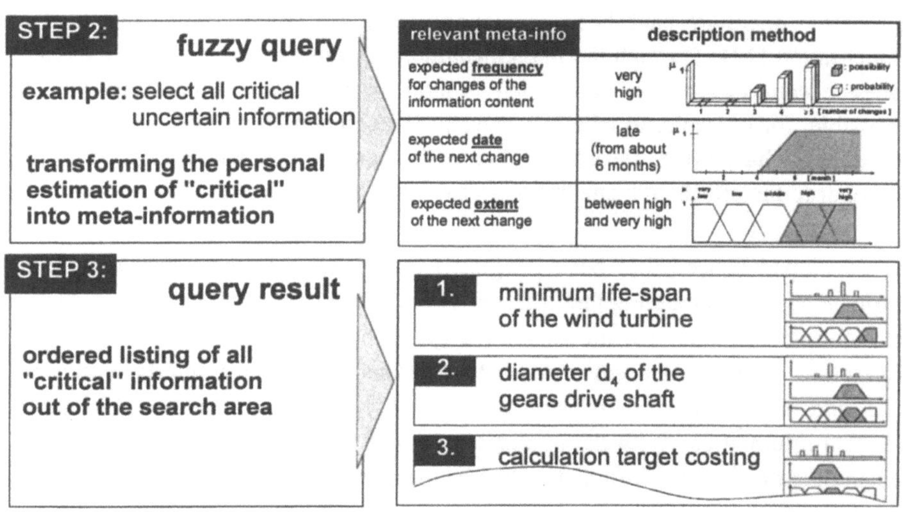

Figure 6 Second and third step for the use of information-maturity in Simultaneous Engineering

The application of "fuzzy pattern matching techniques" provides the means to compare the fuzzy query with all uncertain information in the search area [8]. These techniques determine a degree of matching or mismatching of a pattern and a data fuzzy set. The result of the pattern matching process is the input for the third step.

By applying "fuzzy ranking techniques" the selected uncertain information units can be ordered depending on the extent to which they fit the query [1]. Especially if there are huge amounts of selected data, the user may wish the data to be in an order which corresponds to his preferences. It is not sufficient just to list the selected information. With the application of fuzzy ranking techniques it is possible to rank the answers contained in the database according to the degree to which they satisfy the query.

The application of fuzzy query techniques offers a more flexible and user-friendly access to uncertain information. But fuzzy set theory contains further potentials to enrich query functionalities.

This especially includes features which allow a more intuitive information retrieval. Conventional data base operation is based on Boolean operations. Human information processing usually does not correspond identically with these operations. The semantic meaning of 'AND' usually is not equivalent to the logical 'AND'. In nearly any human selection process one will find the situation that one criterion compensates another one to a certain degree. In fuzzy set theory compensatory aggregation operators have been developed, which are able to model this kind of compensation [12]. Through the implementation of compensatory queries the query process becomes even more appropriate to intuitive information processing. Thus, it is possible, for example, to take into account the compensation between a date and the extent of change (e.g., the effects of an early changing date with a high extent can be similar to the effects of a late but small extent of information).

The developed concept has been integrated in an object-oriented data base. This prototype supports modeling the different views of the stored information and takes the requirements into account which are given by the dynamic development of information. The product and process descriptions consider the interface standard STEP and have been extended with the capability to model uncertain information and meta-information. With the implementation of the system it can be shown that the handling of uncertain information can be managed.

8 CONCLUSION

The methodology presented supports the semantic extension of existing concepts with the aspect of uncertainty. Based on the interface standard STEP an easy to add model has been developed. The possibility to access meta-information using natural language supports user friendly interface construction. Combined with the introduction of small information units, it is possible to open additional potentials for parallelizing activities in an integrated product and process development. This leads to a shorter and less iterative development lead time.

ACKNOWLEDGMENT

The results presented in this article were worked out in the common research project 361 'Models and Methods for an Integrated Product and Process Development' funded by the German Research Community (DFG) [4].

REFERENCES

[1] Bortolan, G.; Degani, R.: A Review of some Methods for Ranking Fuzzy Subsets in: Readings in Fuzzy Sets for intelligent Systems, Eds.: Dubois, Prade, Yager, 1993

[2] Chen, S.-J.; Hwang, C.-L.: Fuzzy Multiple Attribute Decision Making, Springer, 1992

[3] Dubois, D.; Prade, H.: Possibility Theory: An Approach to Computerized Processing of Uncertainty, New York, London, 1988

[4] Eversheim, W.; Bochtler, W.; Laufenberg, L.: Methods and Models for Integrated Modelling of Product and Processes, Production Engineering, Vol. I/2 (1994), S. 173-176

[5] Eversheim, W.; König, W.; Pfeifer, T.; Weck, M.: Manufacturing Excellence: The Competitive Edge, Chapman and Hall, London, 1994

[6] Eversheim,W., et. al.: Simultaneous Engineering, Erfahrungen aus der Industrie für die Industrie, Springer Verlag, Heidelberg, 1995

[7] Pallischek, O.; Wichhardt, F.: Objektorientierte Unternehmensmodellierung für CIM, CIM-Management, Nr. 4, S. 21-25, 1992

[8] Prade, H.: A Two-Layer Fuzzy Pattern Matching Procedure for the Evaluation of Conditions Involving Vague Quantifiers, Journal of Intelligent and Robotic Systems 3, Kluwer Academic Publishers, S. 93-100, 1990

[9] Schenck, D.; Wilson, P.: Information Modeling the EXPRESS Way, Oxford University Press, Oxford, 1994

[10] Wheelwright, S.-C.; Clark, K.-B.: Revolution der Produktentwicklung: Spitzenleistungen in Schnelligkeit, Effizienz und Qualität durch dynamische Teams, Campus-Verlag, Aachen, 1994

[11] Wildemann, H.: Optimierung von Entwicklungszeiten: Just-in-Time in Forschung & Entwicklung und Konstruktion, Transfer-Centrum-Verlag, München, 1993

[12] Zimmermann, H.-J.: Fuzzy Set Theory, and its Applications, Boston, Dordrecht, Lancaster, 1991

[13] Zimmermann, H.-J.: Fuzzy Technologien, Prinzipien, Werkzeuge, Potentiale, Düsseldorf, 1993

5

APPLICATION OF THE FUZZY AHP METHOD FOR ASSESSING ALTERNATIVE PRODUCTION CYCLES

Professor Dr.-Ing. Dr.-Ing. E.h. Manfred Weck*
Professor Dr.-Ing. Fritz Klocke**
Dipl.-Ing. Hanno Schell*
Dipl.-Ing. Elmar Rüenauver**

Laboratory for Machine Tools and Production Engineering,
** Chair of Machine Tools,*
*** Chair of Manufacturing Technology,*
RWTH Aachen, University of Technology, 52056 Aachen, Germany

ABSTRACT

Production technicians can generally choose any one of a number of alternatives for producing a component. The combination of the possible alternatives leads, in turn, to a large number of production cycles. The production technician now has to decide which of the many production cycles available are best suited to yielding the desired processing results, in line with predefined targets. These targets may include low production costs, an optimum degree of capacity utilisation, minimising environmental pollution, or often similar things. The problem arising, i.e. which alternative to choose, can be tackled with the aid of the Analytical Hierarchical Process (AHP), which stems from the multi-criteria analysis. Due to the integration of product and process design the decision of the production cycle should be made at any early stage of product development. This makes the decision process more difficult. Therefore, the input data loses precision and this must be considered accordingly. The method presented here of evaluating different production cycle alternatives takes due account of this by adding the mathematics of fuzzy logic to the classical AHP. Any production cycle evaluated in this manner yields a fuzzy set. The outcome of the analysis can finally be defuzzified by forming the surface centre of gravity of any fuzzy set, and the alternative production cycles investigated can be ranked in order in terms of the main objective set.

1 INTRODUCTION

The target behind assessing alternative production cycles is to pinpoint connections between the design of the products and possible production alternatives at an early stage in the product design. This establishes a basis for defining the process parameters at an early stage (see **Fig. 1**) /SFB92/. The outcome of the design work should be improved by concurrently considering the production requirements.

With regard to the requirements resulting from process planning and early technological planning, the connections between the product design and possible production alternatives need to be described. Taking that as a basis, alternative production technologies were investigated in order to establish just how suitable they were for the production of a product; they were assessed on the basis of various target criteria and defined areas of priority. Multi-criteria analysis techniques were used for this purpose; they had to be analysed and adapted to suit the problems of assessing production cycles.

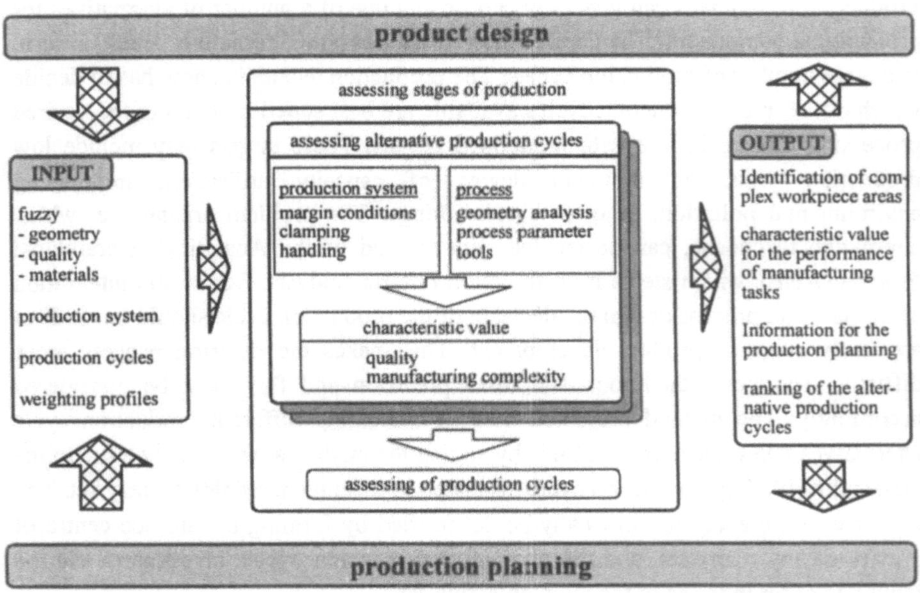

Figure 1: Target in assessing alternative production cycles

A system of characteristic values was developed for the assessment procedure, allowing us to characterise, assign a weight to and assess individual stages of production /SFB95a/. The various stages of production were investigated in terms of expenditure, reliability and the tools required.

In order to incorporate this evaluation procedure into product development at an early stage, it is essential that unreliable data can be processed. The available information relating to the product and to the probable production process might be incomplete or partly undefined. The assessment procedure must be engineered in such a manner that production cycles can be assessed even before product design work is completed. At the same time this means that the fuzziness of the input variables is reflected in the outcome of the assessment procedure.

Due to the variety of different criteria derived from a number of different production processes, a system of characteristic values was employed. So it is possible to make an assessment in all by taking due account of all factors of relevance to production. When listing the criteria upon which to base the assessment, individual stages of production and handling had to be taken into account, as had the analysis of complete production cycles.

2 SELECTING WHICH ASSESSMENT METHOD TO APPLY

The assessment method used must enable us to choose the one production alternative in a finite selection which is best suited in terms of the overall target. Various priorities and part-targets should be allowed for in defining the overall target, such as minimising production costs, reliable adherence to the required quality standards, minimum investment costs and reliability in terms of the alternatives referred to for the purposes of the assessment. The assessment must not only take the intended targets into consideration, it also has to allow for a large number of different criteria, which are included in the decision-making process by devising a system of characteristic values. The resulting complexity of the assessment task calls for an approach which makes it possible to apply a weighting profile to the individual part-targets. At the same time, very different criteria are taken into consideration. The assessment procedure is made all the more difficult by the fuzzy weighting of the targets and the fuzzy input data (see **Fig. 2**). The assessment methods and procedures in question are combined under the common heading 'multi-criteria analysis' /Lau82, Zim91/.

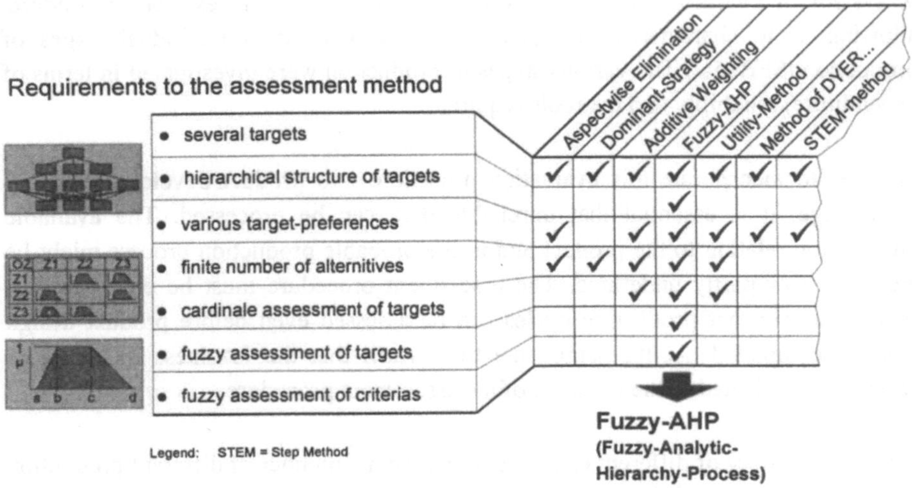

Figure 2: Structuring the decision-making problem and deriving a suitable method of assessment

MADM (multi-attribute decision-making) procedures are applied for tackling this decision-making problem. In these techniques, the array of alternatives must be predefined. This grouping comprises the production alternatives which are explicitly known and which thus form a discrete set for solving the problem. Each alternative is evaluated on its attributes (criteria and targets), which do not necessarily have to be described in figures. The decision-making problem is solved by comparing the attributes and defining points of merit (key points) of different alternatives relating to specific attributes. Whenever MADM procedures are applied, searching for an optimum decision-making procedure is very much in the forefront and the considered alternatives will be ranked. One application of this method is known as the analytic hierarchy process (AHP), which provides a simple way of analysing problematic situations and preparing the way for taking decisions /Saa87/.

In order to apply the AHP, the decision-making problem must be structured into different hierarchical levels. A precise overall target exists for every decision which has to be taken. In our case, this involves choosing the best production cycle in terms of sub-targets, i.e. costs, time, quality, environmental compatibility, etc. The sub-targets, in turn, are described by further criteria which lie one stage lower in the hierarchical structure. For example, the sub-target time comprises the following criteria: machining time, non-productive time, transport time and idle time (**Fig. 3**).

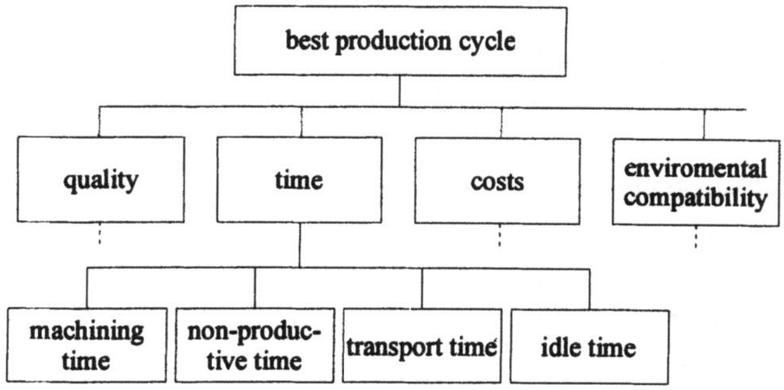

Figure 3: Extract of the hierarchy in a decision-making problem

Fuzzy data needs to be processed before the assessment technique can be applied in an early phase of product design (Fig. 2). Extending the AHP method to include the possibility of expressing the assessments and the weightings between the individual criteria in fuzzy sets safeguards the use of this unreliable and fuzzy data /Laa83, Buc84/. Therefore, the fuzzy AHP method was chosen as a suitable assessment procedure to apply here. This method was adapted to suit the defined task, i.e. assessing alternative production processes. One measure which was particularly important here was introducing K_v, a characteristic number relating to specific processes. This will be described in detail later on.

The fuzzy AHP method, or the AHP in general, is essentially based on a comparison between pairs of alternatives from the finite set A of all of the alternatives. The decision-maker must be able to define a value $a_{ij} \neq 0$, to any two alternatives i and j, for the purposes of comparing these two alternatives with regard to the main criteria. The following would thus apply:

$$a_{ij} = \frac{1}{a_{ji}} \qquad \text{for all } i, j \in A \qquad (1)$$

This requirement stipulates that each matrix of comparisons between pairs is reciprocal. If the decision-maker believes that alternative i is 5 times more important than alternative j, then alternative j must only have one-fifth the degree of importance of alternative i. One consequence of this reciprocal relationship is that if pairs are compared for a total of n alternatives, n*(n-1)/2 comparisons actually need to be performed in order to complete the matrix of comparisons between pairs $(a_{ij})_{1 \leq (i, j) \leq n}$. This is a simplification for determining the required matrices and for establishing a computerised system of assessment. However, it must be checked accordingly in order to prevent there being any inconsistencies in the information.

A significant advantage of the Buckley fuzzy AHP method is that these relations can also be described by using linguistic expressions, such as "roughly 3 to 1" or "about 5 to 1" in terms of fuzzy sets. The fuzzy sets are trapezoids formulated as follows:- $(\alpha_{ij}, \beta_{ij}, \gamma_{ij}, \delta_{ij})$, whereby the following applies: $0 \leq \alpha_{ij} \leq \beta_{ij} \leq \gamma_{ij} \leq \delta_{ij}$.

Another advantage of the Buckley fuzzy AHP method is that inconsistencies may also arise in comparisons between pairs, which is unavoidable in this assessment task in view of the large number of criteria which have to be compared.

3 METHODS OF WEIGHTING AND ASSESSING STAGES OF PRODUCTION

The devised concept was transformed into a system of characteristic values and provides the assessment of individual stages within a production cycle and the assessment of production cycle as an inter-related whole. In attempting to find a solution, the decision-maker must assign a value W to each assessment criterion. Every assessment criterion includes criteria which are common to a number of different processes and criteria which are specific to one particular process. W reflects the significance of the criterion of one particular stage of production in terms of the assessment targets. These values are derived from a qualitative evaluation and are accorded on a useful-value analysis scale of between 1 and 10 /Zan70/.

Attention must also be paid to the fact that a stage of production with a lower level of process reliability is to be assessed with a lower value at the end of the cycle than at the start. This is because of the capital which has already been committed. The relevance of this stage of production is assigned a weighting factor G, which can be valued at between 0.1 and 1. As the reliability factor diminishes with each additional stage within a production cycle, weighting factor G normally becomes smaller and smaller. The fuzziness and unreliability of the product-defining data available during an early stage is accounted for by fuzzy factor U; it also has a value of between 0.1 and 1.

Multiplying the significance W by the weighting factor G and by the fuzziness factor U yields a characteristic number

$$K_m = W_m \cdot G_m \cdot U_m, \tag{2}$$

where m denotes the individual criteria.

In view of the comparisons which have to be drawn between pairs, it is essential when applying the fuzzy AHP method described above that identical assessment criteria apply to each of the different production cycle alternatives.

Given the fact that there may be some differences in the criteria which have to be taken into consideration in this assessment of the individual stages of production (such as machinability during machine-cutting processes and plasticity during deforming), a variety of different assessment criteria exist. In order to overcome this difficulty, the criteria which cannot be applied to all of the production processes are classed as K_v, a number which is specific to particular processes. This characteristic number is established for every alternative to be assessed so that a comparison of pairs is ensured.

In order to establish this number which is specific to particular processes, K_v values are assigned to the various criteria relating to each individual stage in a production cycle. This results in a matrix of numbers for this production cycle, formulated as $K_{m,n}$, where n represents the number of stages in this production cycle. An aggregate relative number K_n, representing one particular stage in the production process, is calculated from the individual numbers $K_{m,n}$ in this matrix. This aggregate is calculated as follows:

$$K_n = \frac{\sum_{i=1}^{m} \frac{K_{i,n}}{\max K_{i,n}}}{m} \cdot \max K_n \tag{3}$$

where: K_n = aggregate number for a given stage in a production cycle

$K_{m,n}$ = number of a criterion within a stage of production

$\max K_{i,n}$ = maximum possible value of numbers $K_{i,n}$

$\max K_n$ = maximum possible value of numbers K_n

m = criterion number

n = production stage number

This ensures that K_n always falls within 1 and 10 and can be referred back to the same scale of assessment.

The number values calculated in this manner are then converted into a specific process number, applying the following formula:

$$K_v = \frac{\sum_{i=1}^{n} \frac{K_i}{\max K_i}}{n} \cdot \max K_v \qquad (4)$$

where: K_v = number which is specific to a particular process
$\max K_i$ = maximum possible value of numbers K_i
$\max K_v$ = maximum possible value of numbers K_v
n = production stage number

The procedure, described above, for devising the specific number for a particular process is illustrated graphically in **Fig. 4** using sample numbers.

criteria m	value assignment $W_{m,n}$					weighting factor G_n					fuzziness factor U_n					charact. number $K_{m,n}=W_{m,n} \cdot G_n \cdot U_n$				
	stages of production					stages of production					stages of production					stages of production				
	1	2	3	4	5	1	2	3	4	5	1	2	3	4	5	1	2	3	4	5
1	8	8	7	8	8											6,4	5,8	4,5	4,5	3,8
2	9	9	9	9	9	1	0,9	0,8	0,7	0,6	0,8	0,8	0,8	0,8	0,8	7,2	6,5	5,8	5,0	4,3
3	8	9	9	9	8											6,4	6,5	5,8	5,0	3,8
4	6	8	7	8	8											4,8	5,8	4,5	4,5	3,8

$\sum_{i=1}^{m} K_{m,n}$	24,8	24,6	24,6	19,0	15,7
transmission in K_n	6,2	6,2	6,2	4,8	3,9

$\sum_{i=1}^{n} K_n$	27,3
transmission in K_v	5,5

specific number to a particular process
K_v=5,5

Figure 4: Determining K_v, the number relating to a specific process

The example illustrated here is an extract from a system of devising numbers for the production of a gear shaft (cf. Fig. 6 - alternative A3), described later on. A value is assigned to each of the different stages of production (i.e deforming (1), friction welding (2), tempering (3), hard turning (4), grinding (5)) for the various assessment criteria (m) (for example, quantity (1), degree of automation (2), flexibility (3), process reliability (4)). If these values are multiplied by the weighting factors G_n and the fuzziness factors U_n, this then indicates the numbers for the different criteria m for each stage of production n. Applying the calculation algorithm described above, the number relating to a specific process K_v, which is required for applying the fuzzy AHP method, can be deduced from the afore-mentioned numbers. K_v encompasses the influence of all of the assessment criteria in each of the individual stages of production. The advantage of the procedure selected is that production cycles with different numbers of production stages and different production processes, necessitating different evaluation criteria, can be assessed and compared.

4 ASSESSING PRODUCTION CYCLES

Fig. 5 illustrates the general approach to applying the Buckley fuzzy AHP method. First of all, a weighting vector has to be derived from the matrices relating to comparisons between pairs, using the Buckley method, which corresponds to a system of geometrical averaging. This weighting vector reflects the ratios of the assessment criteria in terms of the assessment targets. On the other hand, the criteria in each of the stages for each of the production alternatives have to be assigned values. The number relating to a specific process is derived from some of these criteria, as described above. The specific criteria vector for the production cycle is derived from the number relating to a specific process (K_v) on the basis of other numbers which are specific to the production cycle in question. This criteria vector, which exists for each production cycle, is then fuzzy-multiplied by the weighting vector derived previously. This multiplication procedure yields the weighted numbers for each alternative production cycle. The sum total of these weighted numbers for each of the production cycles amounts to the fuzzy set of results. The alternative production cycles can now be arranged in order of rank by applying a suitable process of defuzzification.

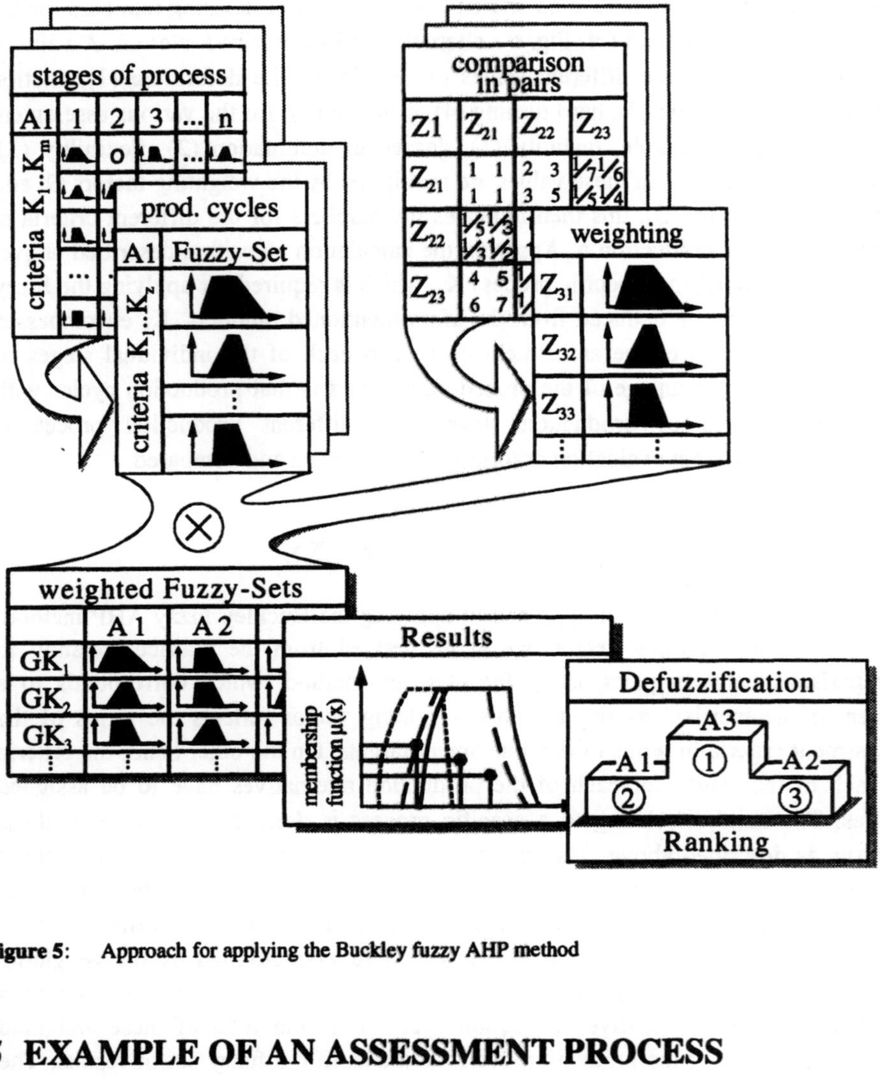

Figure 5: Approach for applying the Buckley fuzzy AHP method

5 EXAMPLE OF AN ASSESSMENT PROCESS

The effectiveness of the selected assessment technique can be illustrated using a case study. A gear shaft has been selected for this, being manufactured conventionally as a full shaft involving one of two possible production cycles and as a sleeve shaft. **Fig. 6** shows the gear shafts and the production procedures in the production cycles, which are to be accorded different values. In alternative A1, the contour of the shaft is turned in the soft material. The toothed wheel mechanism is then milled, and the shaft is case-hardened. The last stage of the process involves grinding the toothed gear mechanism and the seats of the bearings.

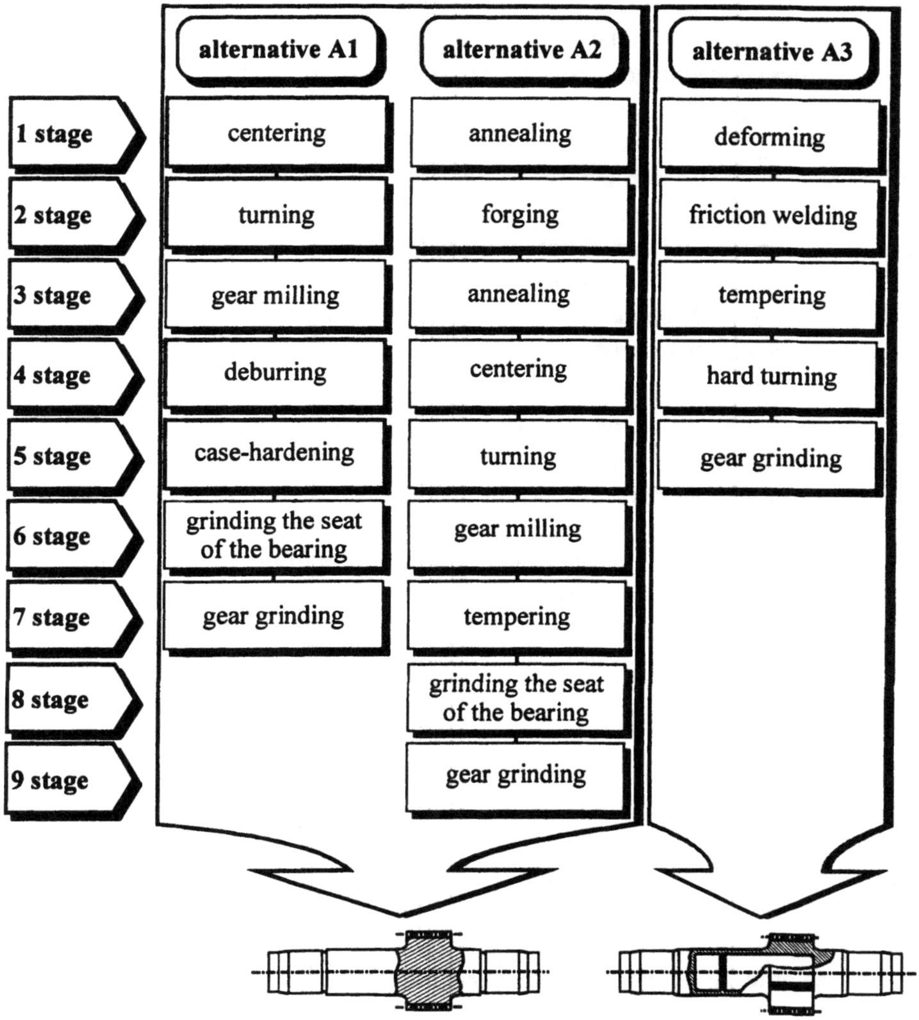

Figure 6: Cycles for manufacturing a gear shaft

In alternative A2, a section of rod is reshaped in four stages after soft-annealing. A grain texture which is suitable for machining is generated in a second thermal treatment process. The contour is also created by the defined machining process during this production cycle before case-hardening and the subsequent grinding work. The gear shaft in alternative A3 comprises two reshaped shaft halves, which are joined together by a process of friction welding. After hardening, the end contour of the bearing seats is created with the required degree of precision, employing a process of hard machining with a defined cutting edge. The finishing grinding work is then performed on the toothed gear mechanisms.

We are now faced with the task of deciding which of three production alternatives best meets our defined targets, using the weighting profile for the assessment criteria. Before we can apply the fuzzy AHP method described above, we first of all need to define the weighting factors for each of the individual assessment criteria. Numbers have to be derived by according values. Values are assigned to each of the criteria, in the individual stages of production, which are of any relevance to the assessment process. The criteria are then assessed which characterise the production cycle as an inter-connected entity. The fuzzy AHP method is based on forming comparisons between the pairs of alternatives being assessed. Therefore, numbers have to be derived for each of the alternatives.

The result of the assessment performed is illustrated in **Fig. 7** in the form of calculated fuzzy sets. These fuzzy sets establish a connection between rating x of the alternative and its membership function $\mu_U(x)$, which indicates the unreliability of the assessment result. It is clearly evident that the fuzzy sets for the different alternatives differ in terms of their ratings and in terms of the plots of their curves. The differences are marginal when comparing alternatives A1 and A2, which describe the production cycles for manufacturing the conventional full gear shafts. However, the assessment result for the production cycle for the modified gear shaft contrasts significantly with these.

The decision-maker will be able carry out direct comparisons between the alternatives once a suitable process of defuzzyfication has been implemented, for example by defining the surface center of gravity of the individual fuzzy sets. The criterion for selecting the most suitable production cycle is that it has as high an alternative rating x as possible, combined with a surface center of gravity $P\mu_U(A)$ with the highest possible membership function $\mu_U(A)$ value. This corresponds to a low degree of fuzziness, coupled with a very reliable assessment result.

If alternatives are being assessed which have identical membership function values in terms of surface center of gravity, the selection should be made on the basis of whichever alternative has a higher rating. In this example, alternative A2 satisfies the requirements better than alternative A1. If the surface centers of gravity differ both in terms of rating and with regard to the value of the membership function, the results must be interpreted.

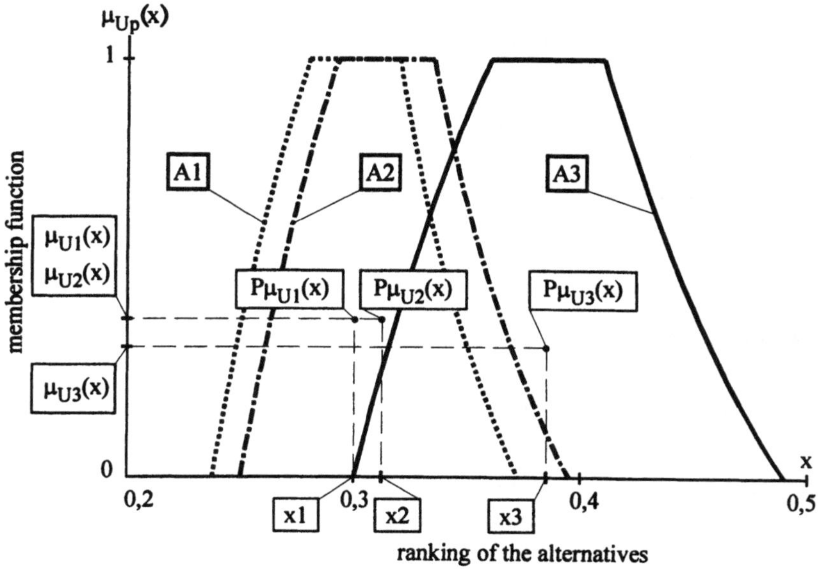

Figure 7: Result of the assessment of the alternatives

The surface center of gravity $P\mu_{U3}(x)$ relating to alternative A3 has a lower membership function value $\mu_U(x)$ than the corresponding surface center of gravity for alternatives A1 and A2. This is derived from the fact that the slopes of the fuzzy set are flatter. The lower the membership function value, the wider the area; correspondingly, the surface center of gravity has a lower membership value. In this example we now have to decide what our priorities are when it comes to interpreting the results.

A high rating in the assessment of the alternatives guarantees that the alternative in question comes very close to meeting the requirements laid down. However, if the alternative's membership function value is lower than that of the other alternatives, this would indicate that it is more fuzzy, thus rendering the result more unreliable. We have to weigh up the assessed rating of the alternative, defined using the fuzzy AHP method, against the membership function value. In the case study, production cycle A3 is clearly preferred as it has a significantly higher rating, coupled with a marginally greater degree of unreliability and fuzziness. A more accurate method and approach for interpreting these results has yet to be developed and will be examined in more in-depth research. /SFB95b/.

ACKNOWLEDGEMENT

This paper is based on the results of a research project funded by the Deutsche Forschungsgemeinschaft (German Research Association) "Models and Methods for the Integrated Modelling of Products and Processes"

REFERENCES

/Buc84/ Buckley, J.J.: The Multiple-Judge, Multiple Criteria Ranking Problem: A Fuzzy-Set Approach, Fuzzy Sets and Systems, Vol. 13, 1984, pp. 25 - 38

/Cha83/ Chankong, V. H., Yacov, Y.: Multiobjective Decision Making - Theory and Methodology, Verlag North Holland, New York, Amsterdam, Oxford, 1983

/Laa83/ Laarhoven, P.J.M., Pedrycz, W.: A Fuzzy extension of Saaty's priority theory, Fuzzy Sets and Systems, Vol. 11, No. 3, 1983, pp. 229 - 241

/Lau82/ Laux, H.: Entscheidungstheorie, Band 1 Grundlagen, Springer Verlag, Berlin, Heidelberg, New York, 1982

/Saa87/ Saaty, R.W.: Tha Analytic Hierarchy Process - What It Is and How It Is Used; Mathematically Modelling, Volume 9, No 3-5, 1987, pp. 161 - 176

/SFB92/ SFB 1427: Modelle und Methoden zur parallelen Produkt- und Prozeßgestaltung, FInanzierungsantrag (Teilprojekt B3), 1992 (Eigendruck RWTH Aachen)

/SFB95a/ SFB 361: Modelle und Methoden zur integrierten Produkt- und Prozeßgestaltung, Abschlußbericht (Teilprojekt B3), 1995 (Eigendruck RWTH Aachen)

/SFB95b/ SFB 361: Modelle und Methoden zur integrierten Produkt- und Prozeßgestaltung, Finanzierungsantrag (Teilprojekt B3), 1995 (Eigendruck RWTH Aachen)

/Zan70/ Zangemeister, Ch.: Nutzwertanalyse in der Systemtechnik, Wittemannsche Buchhandlung, 1970

/Zim91/ Zimmermann, H. J., Gutsche, L.: Multi-Criteria-Analyse, Springer Verlag, 1991

6

A METHOD FOR PERSONNEL SELECTION IN CONCURRENT ENGINEERING USING FUZZY SETS

Dipl.-Ing. Jürgen Stahl

Institute of Industrial Engineering and Ergonomics
RWTH Aachen, University of Technology, 52056 Aachen, Germany

ABSTRACT

In conventional product and process design personnel selection occurs late in the planning process. At that time it is too late to make changes to the production system or initiate other than short-term educational measures. Long-term strategic development of personnel is not possible. Concurrent Engineering (CE) activities start early and are executed parallel to others. This creates a large number of small feedback loops instead of a few large loops and allows requirements of later stages to be taken into account early in the design process. However, methods of CE have to be able to work on the basis of imprecise and poorly defined data. This chapter presents such a method for personnel selection that applies fuzzy set theory to define job requirements and personal attributes of candidates.

1 INTRODUCTION

The role of fuzzy sets in personnel selection in Concurrent Engineering is twofold. First, CE in general requires the exchange of poorly defined information to start activities early and concurrently. Second, with personnel selection it is additionally necessary to describe abilities, skills, etc. which are often not sharply definable. These is the type of imprecision fuzzy sets were developed to define (ZADEH 93).

A research group devoted to *Models and Methods for the Integrated and Concurrent Product and Process Design* and funded by the German National Research Foundation is developing methods that apply fuzzy set theory to inter alia selection of manufacturing processes, machine tools, and personnel. Founded in 1993, it currently is stated to continue until 1998. The results presented in this chapter are part of the work carried out in the field "Planning of Production Systems". Thus, the discussion of personnel selection will be limited to this scope.

2 PERSONNEL SELECTION IN CONCURRENT ENGINEERING

The objective of personnel selection is to determine the suitability of candidates for employment. This can only be achieved by a comparison of a candidate's personal qualities and the demands of the job (LEEK 83). The first step in the personnel selection process is to analyze the target job and identify the critical performance requirements. This is done by various job analysis methods. Conventionally, personnel selection is most often done shortly before the start of production (STAUDT 86). This is because at this stage almost all data about the job are known, as are the job requirements. They are then compared with the kinds of knowledge, skills, abilities, and other personal attributes (KSAOs) of the candidates (BORMAN, PETERSON, and RUSSELL 92). Finally a decision for employment is made.

In CE, personnel selection is considered an integrated element of personnel planning and is concurrently executed with other activities of the design of production systems. CE is defined as "a systematic approach to the integrated, concurrent design of products and their related processes, including manufacture and support. This approach is intended to cause the developers, from the outset, to consider all elements of the product life cycle from conception, through disposal, including quality, cost, schedule, and user requirements" (PENNELL and WINNER 89). Hence, CE takes a holistic point of view. The goals of personnel selection are therefore extended. Besides identifying suitable candidates, it has to

- determine the qualification gap to initiate education measures
- suggest modifications to the production system with respect to available candidates.

These additional goals are especially important when a company wishes to make use of already existing resources in terms of production systems and personnel. It is also possible to take into account strategic aspects like the organization's stance regarding equal employment, affirmative action, and strategic personnel development.

3 INTEGRATED PERSONNEL PLANNING IN CE

The concurrent and integrated execution of personnel selection creates strong links to the other activities of facilities design (e.g. material flow, machine tools, buildings). Structured Analysis (McMENAMNIN and PALMER 84) was used to model these links (see LANGEN, KLONARIS, and STAHL 95 for details). It represents the activities and their corresponding data flows in a hierarchy. This method was chosen, because Structured Analysis identifies the source of data and does not use predecessor-successor-relationships that would predetermine a sequential approach. Instead, an activity is rather started when the person who executes it has decided that the data is good enough. Based on this model the concept for the integrated personnel planning in CE was derived.

Parallel to the activities to design the production system the work organization and task structure is developed. This is done by a modeling tool that was developed in a different research project (SCHUMANN 95, WIMMER and LUCZAK 95). It features an integrated job analysis method to evaluate the design in terms of human factors. Currently the modeling tool is being modified to derive the required KSAOs for the various jobs. In the personnel selection process these are compared with the personal KSAOs of the candidates and suitability and qualification gaps are determined. This information is fed back (Figure 1). The design of the production system and its elements is in the early stages rough and job requirements are not well defined. As planning progresses information become more detailed and precise. A method for personnel selection has to be able to deal with this increasing level of detail.

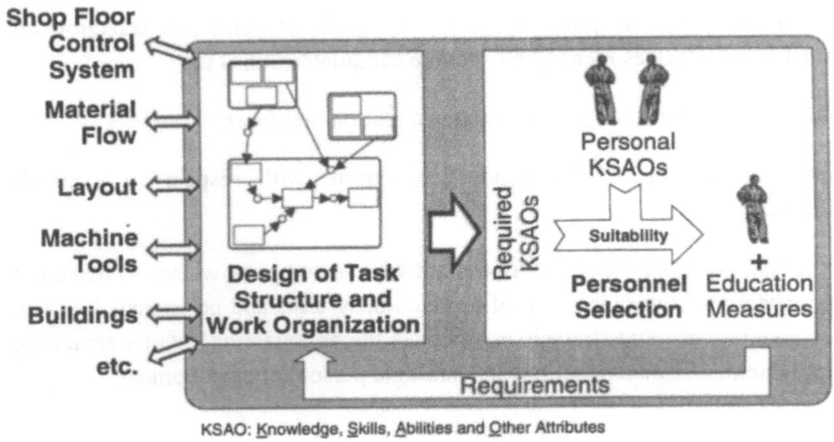

Figure 1: Concept for the Integrated Personnel Planning in CE

4 HIERARCHICAL STRUCTURE OF KSAOs

To describe required and personal KSAOs a hierarchy that comprises their criteria was developed. A small part of its structure can be seen in Figure 2 (for further details see LANGEN, KLONARIS, and STAHL 95). In the beginning of the planning process the upper levels of the structure will be used to describe the job requirements. The more precise the design of the production system becomes, the more criteria from lower levels are added for comparison. The remaining problems are:

1. How is each criterion of the required KSAOs described?

2. How is each criterion of the personal KSAOs described?

3. How are the required and personal KSAOs for each criterion compared and the "suitability per criterion" determined?

4. Finally, how are the suitabilities per criterion summarized to an overall suitability?

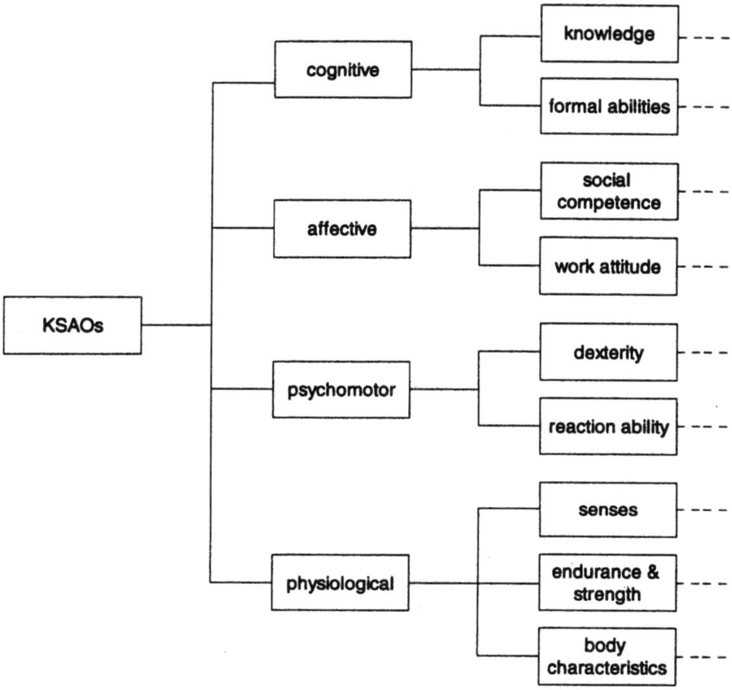

Figure 2: Hierarchical Structure of KSAOs

Considering the KSAO criteria of the hierarchical structure three different types can be identified: ordinal, nominal, and metric. For every type a method of description and comparison must be developed.

4.1 ORDINAL CRITERIA

The values of ordinal criteria (e.g. initiative, motivation, knowledge, etc.) can not be described directly numerically, but can be ordered. Most KSAO criteria are of this type, because humans are most familiar with ordinal values (ZIMMERMANN and GUTSCHE 91).

4.1.1 Description of Personal Ordinal Criteria.

A good way to describe personal ordinal criteria is the use of linguistic variables. These are classically characterized by a quintuple $(x, T(x), U, G, \tilde{M})$, in which x is the name of the variable, $T(x)$ denotes the term set of x, U the universe of discourse,

G a syntactic rule, M a semantic rule for associating with each X its meaning, and $\tilde{M}(X)$ a fuzzy subset of U (ZIMMERMANN 91).

For a proper assignment of fuzzy sets to the terms one has to consider the context of personnel selection. An example may illustrate this situation:

> A welding job requires work on some difficult and some easy workpieces. The best decision will be a person who is able to weld the difficult ones. He will of course be able to handle the easy ones as well.

This example shows that for a comparison the selector will make his decision only with respect to the higher levels of required and personal KSAOs. The low levels can be considered as a "subset" of the higher ones. This leads to the following description of personal ordinal requirement criteria:

- x is the variable for the criterion, e.g. motivation

- $T(x) \subseteq$ {at most none/lowest, at most very low, at most low, at most medium, at most high, at most very high, at most perfect/highest}

  The term set comprises up to 7 levels, that is the maximum number of levels a person can sensibly distinguish (BERNHARD 75). The supplement "at most" is a hedge and takes care that only the high levels are relevant for comparison (Figure 3).

- U=[0,1] is called the *level*. The level of the KSAOs is mapped on the closed interval [0,1]

- $\tilde{M}(T_i)=\left\{\left(u,\mu_{T_i}(u)\right) \mid u \in [0,1]\right\}$

  $\tilde{M}$ assigns membership functions to the terms of T. These may be arbitrary in shape except that $\exists u \in [0,1]$ with $\mu_{T_i}(u)$=1 which means that there should be at least one u where the term fully applies. This condition is not very restrictive as if it were not fulfilled, the term set and the universe of discourse would not quite fit and their definition would have to be questioned.

4.1.2 Description of Required Ordinal Criteria.

The description of required ordinal criteria relies on the modeling tool (see above). The frequency distribution of the activities over the level $h(u)$ is transformed into a membership function $\mu_j(u)$:

$$\mu_j(u)=\begin{cases}\dfrac{h(u)}{H} & \text{for } u>u' \\ 1 & \text{for } u\le u'\end{cases} \quad \text{where } H=\max(\mu) \text{ and } u'=\max\{u\in[0,1]\,|\,\mu_j(u)=H\}$$

which means that (H,u') denotes the rightmost maximum of $h(u)$. Thus, the frequency distribution $h(u)$ is "neglected" for $u \le u'$ and mapped onto the closed interval [0,1] for $u>u'$. This kind of transformation and the use of the hedge "at most" ensure that only the higher levels of personal and required KSAOs are relevant for comparison. Figure 3 gives an example for the membership functions. However, the link between the job activities and the required KSAOs is the subject of current work.

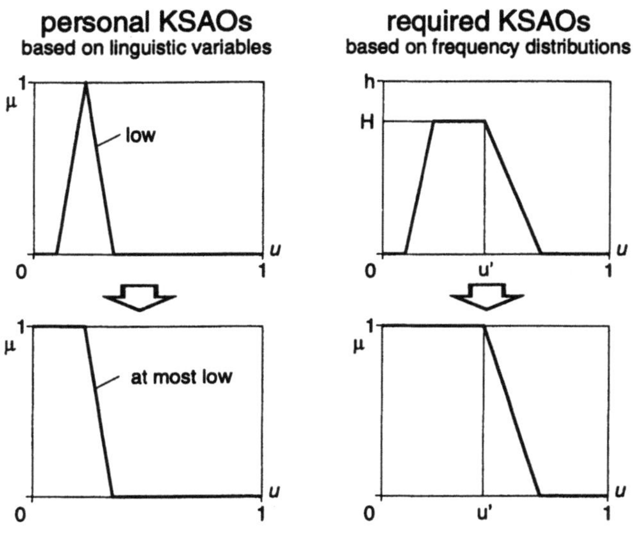

Figure 3: Example for Membership Functions of Ordinal Criteria

4.1.3 Comparison of Ordinal Criteria.

The main precondition for the comparison of personal and required KSAO criteria is the identity of the respective *level* scales. This can be achieved by defining a fixed scale in the task model. Then several scenarios are generated with their frequency distributions and membership functions. These scenarios are rated by the persons who assess the candidates. This way the membership functions for the assessment of personal KSAOs are created on the same basis as the required ones (Figure 4).

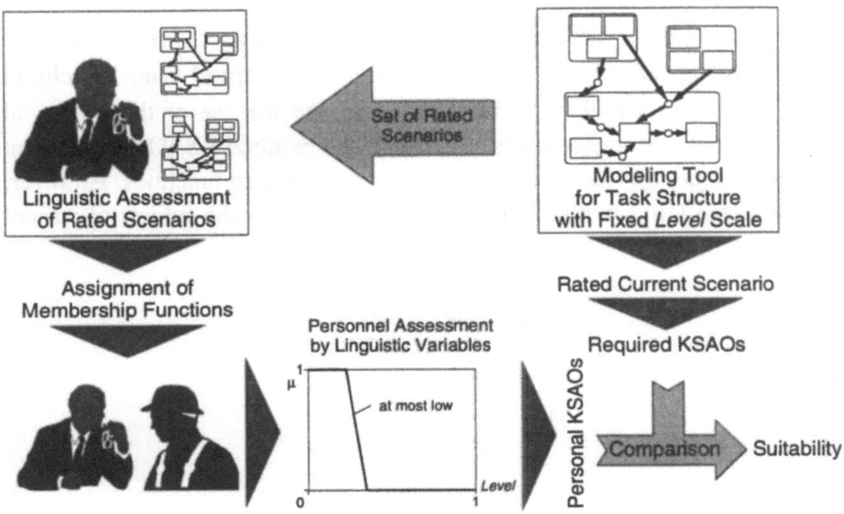

Figure 4: Derivation of the Membership Functions for the Linguistic Description

The goal of the comparison is to determine a suitability number *s* which indicates the match degree of the membership functions μ_p and μ_j and should have the following properties:

1. s=1, if they are identical.

2. s=0, if they are completely different. "Completely different" means that where μ_p=1 (personal KSAO fully applies), μ_j=0 (required KSAO does not apply at all) and vice versa. This is only the case, when a perfect KSAO is required and a candidate does not have the respective KSAO at all or a person has a perfect KSAO, which is not required. This rarely happens in reality.

A suitability number with these properties is

$$s = \int_0^1 1 - \left| \mu_p - \mu_j \right| dx$$

This number is based on the area between the membership functions. As s does not indicate any potential for education and training the qualification deficit per KSAO criterion is expressed separately by the variable d. If it exceeds a certain limit, education measures are necessary. It is defined as

$$d = \frac{1}{2}\left(\int_0^1 \left| \mu_P - \mu_J \right| dx - \int_0^1 \left(\mu_P - \mu_J \right) dx \right)$$

5 NOMINAL CRITERIA

The values of nominal criteria are categories. Every membership of a value in a category must be unique. There is no ranking of categories. Examples for this kind of criteria are sex (male/female), allergies (yes/no), etc.

It is possible to use the nominal values as they are to describe the criteria and Boolean logic to define suitability values. However, to avoid using several different methods we apply the same method for description and comparison as for ordinal ones. They are therefore also described by linguistic variables. This requires the definition of corresponding membership functions on the interval [0,1]. However, the actual value of *level* is in this case artificial. While for ordinal criteria a high value is better than a low one this is not true for nominal variables as there is no ranking. The membership in the categories is then expressed by the use of rounding functions, for example

μ_0=1-ROUND(u) and μ_1=ROUND(u) for a dichotomous criterion

Required nominal KSAOs can not be derived from the modeling tool. They have to be determined manually. Nevertheless, they are defined as personal KSAOs and the comparison is done in the same way as in case of ordinal criteria.

The values of the nominal criteria that are currently in the hierarchy are not affected by education. Therefore an education potential can not be identified and the definition of a qualification deficit d does not make sense. However, if nominal criteria with education potential are introduced in the future, the qualification deficit will be defined as for ordinal criteria.

6 METRIC CRITERIA

The values of metric criteria (e.g. weight, size) consist of a numeric value and a dimension. The personal metric KSAOs can be determined precisely and are therefore described accordingly as a real variable $p \in U$ (e.g. size: 1.87 m, weight: 79 kg). U is the universe of discourse and an interval of possible values u, where u is interpreted as one of the KSAOs, for example age.

The corresponding required KSAOs usually cover a range of values, which may continually be narrowed as the database becomes more detailed. The description of the required metric criteria is done by a fuzzy set J

$$J = \{(x, \mu_J(x)) | x \in U\},$$

where the values of the membership function represent the preference. The suitability value is then simply defined as

$$s = \mu(p).$$

Metric criteria can not be improved by education measures. Therefore the definition of a qualification deficit is not necessary.

7 DETERMINATION OF OVERALL SUITABILITY

The determination of the overall suitability on the basis of a set of suitability values per criterion can be viewed as a Multi Attribute Decision Making problem. The overall suitability can be interpreted as the main goal, the various candidates as alternatives, and the suitability values per criterion and candidate as the judgments with respect to a goal and decision alternative (ZIMMERMANN and GUTSCHE 91). The hierarchical structure of KSAOs is then a hierarchical structure of goals. A well-known method for Multi Attribute Decision Making with a main goal and a structured set of subgoals is AHP (Analytic Hierarchy Process, SAATY 80). AHP allows weights on the goals, which is useful to realistically describe the job requirements.

AHP is based on the compensation of alternative ratings, which means that a poor rating of an alternative with respect to one goal may be compensated by a good rating in another. In personnel selection, however, there may be knock-out criteria that exclude an alternative from further consideration, if a rating exceeds a certain limit. In order to apply AHP it is therefore necessary to consider these criteria beforehand. For the remaining alternatives the overall suitability S_i for each candidate is then calculated as

$$S_i = \sum_j w'_j \cdot s_{ij}$$

where

s_{ij} denote the suitability for the i-th candidate and the j-th criterion,

w'_j denote the weight of the j-th criterion with respect to the main goal.

As the job requirements are not exactly known in the early stages, neither are the weights. It is therefore useful to use fuzzy sets instead of numeric values as weights. This can be done by the use of Fuzzy-AHP (BUCKLEY 85).

The selection of candidates is finally done on the basis of the ranking of the alternatives with respect to overall suitability. This is trivial for AHP, but requires the use of fuzzy ranking methods for Fuzzy-AHP (see BORTOLAN and DEGANI 93 for a review of fuzzy ranking methods). The subjects of education measures for selected candidates are derived from the criteria that are indicated by the qualification deficit variables.

8 CONCLUSION

A method for personnel selection in Concurrent Engineering applying fuzzy sets was presented. Fuzzy sets are well suited for this kind of problem as they can be used to describe both poorly definable and imprecise data. However, it is still necessary to link required KSAO to the task structure and work organization in the modeling tool to derive the frequency distributions automatically. This is one of the objectives of the ongoing work in the research group in the next three years.

REFERENCES

Bernhard U. Wie wird der Mitarbeiter beurteilt? Eine empirische Studie. Personal 1975;2:60-62

Borman, Walter C.; Peterson, Norman G.; Russell, Teresa L. "Selection, Training, and Development of Personnel." In *Handbook of Industrial Engineering*, Gavriel Salvendy, ed. New York: Wiley & Sons, 1992

Bortolan, G.; Degani, R. "A Review of Some Methods for Ranking Fuzzy Subsets" In *Fuzzy Sets for Intelligent Systems*, Didier Dubois, Henri Prade, Ronald R. Yager, eds. San Mateo, CA: Morgan Kaufmann Publishers, 1993

Buckley JJ. Fuzzy Hierarchical Analysis. Fuzzy Sets and Systems 1985;3;233-247

Langen, René; Klonaris, Pavlos; Stahl, Juergen "Einsatzplanung von Produktionsmitteln" In *Modelle und Methoden zur integrierten Produkt- und Prozeßgestaltung*, Report 1993/94, Aachen: SFB-Geschaeftsstelle der RWTH, 1995

Leek, Chris. "Techniques of Manpower Selection and Development." In *Manpower Planning*, Edwards et al., New York: Wiley & Sons, 1983

McMenamin, Stephen. N.; Palmer, John F. *Essential Systems Analysis*. New York: Yourdon Inc., 1984

Pennell J,Winner R Concurrent Engineering: Practices and Prospects. GLOBECOM'89 IEEE Global Telecommunications Conference and Exhibition for the 1990s and Beyond; 1989 November 27-30; Dallas, TX, IEEE (1) 1989

Saaty, Thomas L. The Analytic Hierarchy Process. New York: McGraw-Hill, 1980

Schumann, R. *Entwicklung eines modellgestützten Systems zur Gestaltung teilautonomer Arbeitsgruppen in der Fertigung von Automobilzulieferbetrieben*. Dissertation, Aachen University of Technology, 1995

Staudt, E. "Die Führungsrolle der Personalplanung im technischen Wandel." In *Das Management der Innovationen*, E. Staudt, ed. Frankfurt a. M., 1986

Wimmer R., Luczak H. Modellbildung und Simulation zur prospektiven Aufgabengestaltung in Produktionssystemen. Bericht zum 41. Arbeitswissenschaftlichen Kongreß an der Technischen Universität Wien; 1995 March 29-31; Vienna, Cologne: U. Schmidt, 1995.

Zadeh, Lofti. "Fuzzy Sets." In *Fuzzy Sets for Intelligent Systems*, Didier Dubois, Henri Prade, Ronald R. Yager, eds. San Mateo, CA: Morgan Kaufmann Publishers, 1993

Zimmermann, Hans-Juergen. *Fuzzy Set Theory and its Applications*. Boston: Kluwer Academic Publishers, 1991.

Zimmermann, Hans-Juergen; Lothar Gutsche. *Multi-Criteria Analyse*. Berlin: Springer-Verlag, 1991

INDEX

Activity view, 247
Aggregation, 10, 206, 224
compensatory, 248
operator, 137
restrictions, 12
tree, 111, 114, 225
Air tank design, 3
Aircraft cabin layout, 93, 142, 181, 227–228
Analytic Hierarchy Process (AHP), 228, 254, 274
Annihilation, 13
Artificial intelligence, 152, 226
ASME code, 5
Aspiration level, 135, 208, 222
Assessment method, 253
Attributes, 96, 158, 210, 219
BAB-aspiration level strategy, 140
Backwards mapping, 9, 18
Basic attribute, 202
Basic modules, 164–165, 185
Basic optimization module (BOM), 169, 171
Bellman-Zadeh-Model, 114, 215, 226
Bounded objective method, 57
Brake fade, 30
Branch and bound algorithm, 131, 137, 183, 186
Branch and bound optimization strategy, 138
Building-block, 203
Chen and Hwangs method, 179, 203
Compensation, 225
Compensatory aggregation, 248
Compensatory operator, 111
Components, 92, 96, 131, 156, 202
Composed attributes, 209
Compositional hierarchy, 102, 198
Conceptual constraints, 162, 189, 191, 197
Conceptual hierarchy, 96, 129, 146, 157, 164, 183
Concurrent engineering, 2, 47, 265–266
Configuration, 89–90, 92, 135, 226
model, 116, 118
Constraint language, 184
Constraint-propagation, 189
Constraints, 89, 92, 96, 143, 164–165, 183
Context-information, 0, 238
Control strategy, 129, 185
Control, 165, 183
Core model, 163, 218–219
Cost-of-reducing-emission, 211, 214
Crisp constant, 33
Crisp constraint, 95, 111, 114, 131, 212
Criteria
metric, 274
nominal, 273
ordinal, 269
Data-base, 177
Decision analysis, 155
Decomposition methods, 123
Defuzzification, 111–112, 179, 207, 226, 259, 262

Deposition, 211, 214
Design, 89, 91–92, 226
  alternatives, 10
  preferences, 8
  variable space (DVS), 3
  variables, 3, 6, 32, 34
Development lead time, 233
Document view, 247
Domain knowledge base, 93, 95, 197, 228
Domain knowledge model, 156, 219–220
Domain objects, 164, 184
Dubois and Prade, 19, 24, 26
Emission, 211, 214
Energy considerations, 43
Engineering design, 90–91
Environment receptors, 157
Environmental research, 90, 155, 210, 227
Evolutionary algorithm, 153, 201, 228
Expert system, 145
EXPRESS, 237
EXPRESS-G, 244
Feedback, 244
Fitness functions, 153, 155
Fitness, 155, 208
FRAME languages, 227
FRAME representation, 100
Frames, 160
Frictional torque, 33
Functional properties, 105
Functional requirements, 168, 210
Fuzzy
  AHP, 196, 255
  attributes, 168
  constraints, 95, 110, 168, 188, 212, 223
  design requirement model, 112, 226
  design requirement, 109, 113
  design-optimization model, 114
  frame hierarchies, 101
  frames, 95, 223
  games, 65
  goals, 106, 110, 168, 181, 189, 201, 215
  IF-THEN-rules, 217
  is-a-hierarchy, 166, 168
  logic, 90, 211, 226
  MADM, 95, 114, 131, 176, 183, 226
  measure, 19
  membership, 6
  MODM, 95
  multi-criteria design model, 113, 115
  multi-criteria fitness, 201
  number, 6, 34, 141, 200
  operators, 224
  Pareto-optimality, 62
  pattern matching, 248
  query, 246, 248
  ranking, 248
  relation, 217
  set, 108–109, 203, 206, 212
Fuzzy
  Weighted Average (FWA), 9
Gamma operator, 206, 224–225
Genetic algorithms, 153
Global constraints, 105, 124, 210
Global criterion method, 55
Goal attainment method, 57
Goal hierarchy, 194, 228
Goal programming, 56, 136
Goals, 89, 168, 178, 187, 210
Graphic Interface System (GIS), 151
Heuristics, 150, 164, 199
Hierarchical knowledge, 226
Hierarchical optimization, 118
Hub, 149, 151, 199
Hybrid approach, 89–90, 131, 140

Hyperlink, 242
Hypertext, 242
Imprecise attributes, 89, 95, 100, 173
Imprecise requirements, 210
Imprecision, 2, 32
  intrinsic, 2
Imprecision
  Method of, 2
Induced preference, 9, 33–34, 43
Information unit, 244
Information-content, 235
Information-maturity, 239
Inheritance, 183
Innovative design, 91–92, 100, 226
Intelligent Decision Support Systems (IDSS), 89, 226
Intelligent systems, 89
Is-a-hierarchy, 96, 165
Knowledge base, 147, 178, 220–221
Knowledge model, 115, 163, 183, 190, 218, 226
Knowledge-based configuration, 145, 152, 226
Knowledge-based systems, 90, 152, 226
KONWERK, 95, 145, 164, 183, 216, 227
KSAO, 266
Lebesgue measure, 19, 22
Level Interval Algorithm (LIA), 9
Lexicograhic method, 58
Linguistic terms, 109, 203, 213
Linguistic values, 166
Linguistic variable, 100, 112, 173, 177, 196, 200, 213, 235
LISP, 219
Location, 149
Logistic layout, 90, 149, 197, 227–228
Lottery question, 6
MADM (multi-attribute decision-making), 254
MADM, 114, 170
MCDM Model Generator, 112
MCDM, 219–221
Mechanism design, 71
Membership satiation coefficient, 64
Membership-degrees, 101
Membership-functions, 101, 201, 203, 212
Meta-information, 239
Method of Imprecision, 2
Min-Operator, 101, 141, 176, 205, 217, 224
Misinterpretation, 238
Mixed integer problem, 228
Modelbase, 93
Modelling languages, 156
Modelling support system, 95, 113, 156
MOMIP, 200
Multi-citeria analysis, 252–253
Multi-criteria decision making (MCDM), 90, 163
Multi-criteria, 227
Mutation, 154
Necessity measure, 24
Noise variable space (NVS), 19
Noise variables, 8, 19
Non-linear, 29
Nonlinear membership functions, 64
Object-oriented model, 237
Objectives, 105–106, 124, 210
Optimization model, 199
Optimization module, 170, 186
Optimization, 18, 105, 131, 202, 227
Optimization-based configuration, 118, 170
Organization view, 247

Papalambros and Wilde, 3
Pareto-optimality, 54
Part-of hierarchy, 96
Pay-off period, 234
PC-configuration, 228
Performance specifications, 8
Performance variable space (PVS), 3
Performance variables, 3, 6, 33–34, 39
Personal computer, 152, 227
Personnel planning, 267
Personnel selection, 265–266
PLAKON, 100, 144
Population, 153, 208
Possibility distribution, 241
Possibility measure, 23
Possibility, 215
Precedence relation, 26
Preference, 5
Preferential order, 6
Preliminary design, 144, 189, 191, 227
Preliminary engineering design, 2
  importance, 2
Probability distribution, 241
Probability measure, 21
Probability space, 21
Product life cycles, 234
Product view, 247
Production alternatives, 252
Production cycles, 0, 259
PROKON, 164, 227
Query process, 245
RAINS, 156, 210, 218
Ranking, 129, 141, 178, 196
Recombination, 154
Relations, 157, 164
Requests, 168
Requirement model, 95, 105, 167, 211, 219, 221, 226
Requirement-type-constraints, 143, 167, 210
Reservation levels, 201–202, 222
Restrictions, 143, 168
ROCKET-SYSTEM, 190
Routine design, 92, 99, 226
Saaty's AHP, 193
Screening method, 178
Seat-pitch, 181
Sensitivity measure, 43
Set-based design, 2, 47
Simple additive weighting method, 178
Simulation languages, 156
Simultaneous Engineering, 233
Source regions, 157, 220
Space launch systems, 144, 189, 227–228
Space structure design, 75
STEP, 237
Stochastic uncertainty, 2, 7
Structured Analysis, 267
Structured modelling, 156
Sugeno, 23
  integral, 23
Trade-off
  compensating, 14, 16
  non-compensating, 13, 16, 29
Trade-offs
  formal, 47
Transportation system, 150
Transportation, 149, 197
Uncertain information, 235
Uncertainty integral, 20
Uncertainty
  stochastic, 2
Uncontrollable effects, 19
Utility function method, 55
Utility theory, 6
Variables
  design, 3
  performance, 3

Vector optimization, 118
Vehicle-routing problems, 200
Weighted sum, 10
XKL, 144

MIX
Papier aus verantwortungsvollen Quellen
Paper from responsible sources
FSC® C105338

If you have any concerns about our products,
you can contact us on
ProductSafety@springernature.com

In case Publisher is established outside the EU,
the EU authorized representative is:
Springer Nature Customer Service Center GmbH
Europaplatz 3, 69115 Heidelberg, Germany

Printed by Libri Plureos GmbH
in Hamburg, Germany